BIOLOGY

CLEP* Study Guide

All rights reserved. This Study Guide, Book and Flashcards are protected under the US Copyright Law. No part of this book or study guide or flashcards may be reproduced, distributed or stored in a retrieval system, or transmitted in any form or by any means, electronic, mechanical, photocopying, recording, or otherwise, without the prior written permission of the publisher Breely Crush Publishing, LLC.

© 2019 Breely Crush Publishing, LLC

*CLEP is a registered trademark of the College Entrance Examination Board which does not endorse this book.

971091218143

Copyright ©2003 - 2019, Breely Crush Publishing, LLC.

All rights reserved.

This Study Guide, Book and Flashcards are protected under the US Copyright Law. No part of this publication may be reproduced, distributed or stored in a retrieval system, or transmitted in any form or by any means, electronic, mechanical, photocopying, recording, or otherwise, without the prior written permission of the publisher Breely Crush Publishing, LLC.

Published by Breely Crush Publishing, LLC
10808 River Front Parkway
South Jordan, UT 84095
www.breelycrushpublishing.com

ISBN-10: 1-61433-559-1
ISBN-13: 978-1-61433-559-7

Printed and bound in the United States of America.

*CLEP is a registered trademark of the College Entrance Examination Board which does not endorse this book.

Table of Contents

Chemical Composition of Organisms	1
Cells	3
Enzymes	6
Energy Transformations	8
Cell Division	11
Chemical Nature of the Gene	16
Photosynthesis	20
Food Chain	21
Biology Classifications and Definitions	21
Ecosystem	22
Charles Darwin	23
Genetics	23
Animal Breeding	27
Rh Factor	28
Albinism	28
Tay-Sachs Disease	28
Cystic Fibrosis	29
Sickle Cell Anemia	32
Prenatal Genetics	32
Human Biology	33
Development after Birth	35
Body Essentials	37
Bacteria	37
Disease	38
Functions of the Body	38
The Immune System and Immunizations	40
Temperature Regulation in Animals	40
Insects	41
Plants	41
Flower Reproductive Parts	43
Research	44
Sample Test Questions	51
Test Taking Strategies	98
What Your Score Means	98
Test Preparation	99
Legal Note	99
References	100

 # Chemical Composition of Organisms

SIMPLE CHEMICAL REACTIONS AND BONDS

The basic building blocks of organisms are hydrocarbons, compounds with a backbone of carbons covalently bonded to hydrogens and other carbons. Hydrogens can be replaced by functional groups such as the hydroxyl group (contains hydrogen and oxygen), carbonyl (carbon double-bonded to oxygen), carboxyl (carbon double-bonded to oxygen and single-bonded to a hydroxyl), amino (nitrogen bonded to two hydrogens), sulfhydryl (carbon bonded to sulfur) or phosphate (phosphorus single-bonded to three oxygens and double-bonded to one oxygen). To make larger molecules called polymers, a hydrogen (H) and a hydroxyl (-OH) group are removed and combined with each other to form a water molecule (H_2O) in a dehydration synthesis reaction.

CHEMICAL STRUCTURE OF CARBOHYDRATES, LIPIDS, PROTEINS AND NUCLEIC ACIDS

The four basic groups of polymers or macromolecules are carbohydrates, lipids, proteins and nucleic acids. Carbohydrates, many of which are sugars, contain carbon, hydrogen and oxygen in the ratio of 1:2:1. The basic empirical formula for carbohydrates is $(CH_2O)_n$ where n is the number of carbon atoms. Energy is released when a C-H bond is broken. Glucose, a six-carbon sugar in a ring formation, has seven energy-storing C-H bonds. Many glucose polymers found in plants are starches. Glycogen is the animal version of starch. Carbohydrates such as cellulose (in plants) and chitin (in arthropods) are chains of sugars which resist digestion because most organisms lack the needed digestive enzymes.

Lipids are a group of molecules such as fats and oils that are insoluble in water. Phospholipid molecules can be thought of as having a polar "head" at one end, comprised of the phosphate group and two long non-polar "tails" at the other end. The tails of such molecules tend to form two rows facing each other, forming the basic framework of biological membranes. Saturated fats only have single bonds between carbons, while unsaturated fats have at least one double or triple bond between carbon atoms.

The shape of a protein determines its function. All proteins consist of various sequences of the twenty possible amino acids. An amino acid contains an amino group (NH_2), a carboxyl group (COOH) and a hydrogen atom bonded to a central carbon atom. Protein structure has six levels: (1) the amino acid sequence, (2) coils and sheets, (3) folds or creases called motifs, (4) the three-dimensional shape, (5) functional units called domains, and (6) individual polypeptide subunits. Proteins facilitate seven basic functions in organisms:

- Enzyme catalysis – facilitate chemical reactions in the organism
- Defense – form the basis of the immune system
- Transport – carry O_2 to tissues and CO_2 away from tissues
- Support – maintain the shape and protection of the organism (skin, ligaments, bones, hair)
- Motion – contract muscles
- Regulation – act as messengers to and from brain, muscles, various systems
- Storage – bind calcium and iron for later use

The biochemical activities of cells depend on many proteins which have different structures. Cells produce those proteins using the information passed from one generation to the next in nucleic acids of DNA and RNA. DNA (deoxyribonucleic acid) encodes the information used to assemble proteins while RNA (ribonucleic acid) "reads" the DNA-encoded information and directs the protein synthesis. DNA is a double-stranded helix, but RNA is a single-stranded molecule.

PROPERTIES OF WATER

Water has a variety of properties that make it a near-perfect molecule for life. Its most outstanding chemical property is that it forms weak chemical associations known as hydrogen bonds which are responsible for many of its physical properties. This happens because its shape gives it a partial negative charge on the oxygen end and a partial positive charge on the hydrogen end. This polarity allows it to attract other water molecules so that it is cohesive and to attract other polar molecules, creating adhesion. Its cohesiveness contributes to the fact that it is liquid at a relative low temperature. Its high specific heat allows it to store vast amounts of energy in the form of heat. The cohesion of water is responsible for its surface tension. Water's high heat of vaporization allows the surface of a human or animal to cool when sweat evaporates. Water can moderate high temperatures through its high specific heat and its high heat of vaporization. Water is an effective solvent for any molecule having a positive and a negative end. Water is also a very effective buffer in a chemical reaction.

ORIGIN OF LIFE

There are three main theories on the origin of life, each with various sub-theories. The main theories are called abiogenesis, cosmozoic and Supreme Being. The abiogenesis theory implies that life has come from some form of non-life. It includes the sub-theories of spontaneous generation, evolution, phylogenesis and the big bang theory. Scientists who believe one of the sub-theories of abiogenesis back their thinking up with thoughts of how life started by the right set of chemicals getting together in the right sequence at the right time in the oceans, in a soup of chemicals, or out of clay and then evolved into worms, then fish, then mammals, and finally man. Those who believe cosmozoic theories have various philosophies as to how man came from space. Some suspect that the building blocks of life came from the cosmos, while others think a lower form of life was dropped off on earth. Believers of the Supreme Being theories include Christians who believe God created Adam from clay and his wife Eve from one of Adam's ribs, as well as Buddhists who believe Buddha created the first man.

Cells

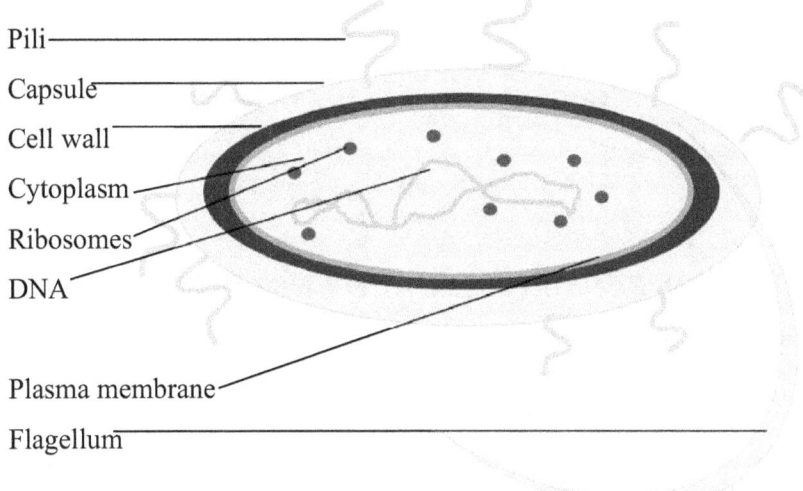

Prokaryotic Cell

- Pili
- Capsule
- Cell wall
- Cytoplasm
- Ribosomes
- DNA
- Plasma membrane
- Flagellum

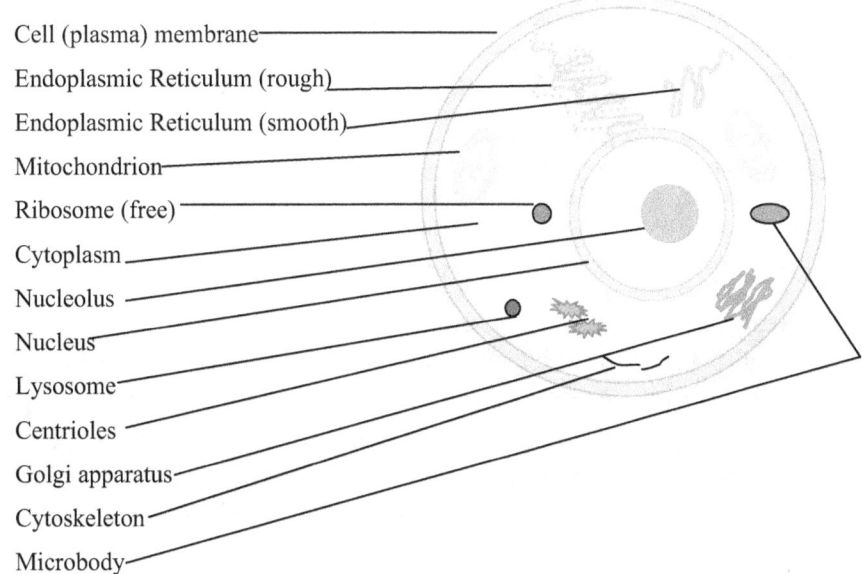

Generalized Eurkaryotic (Animal) Cell

- Cell (plasma) membrane
- Endoplasmic Reticulum (rough)
- Endoplasmic Reticulum (smooth)
- Mitochondrion
- Ribosome (free)
- Cytoplasm
- Nucleolus
- Nucleus
- Lysosome
- Centrioles
- Golgi apparatus
- Cytoskeleton
- Microbody

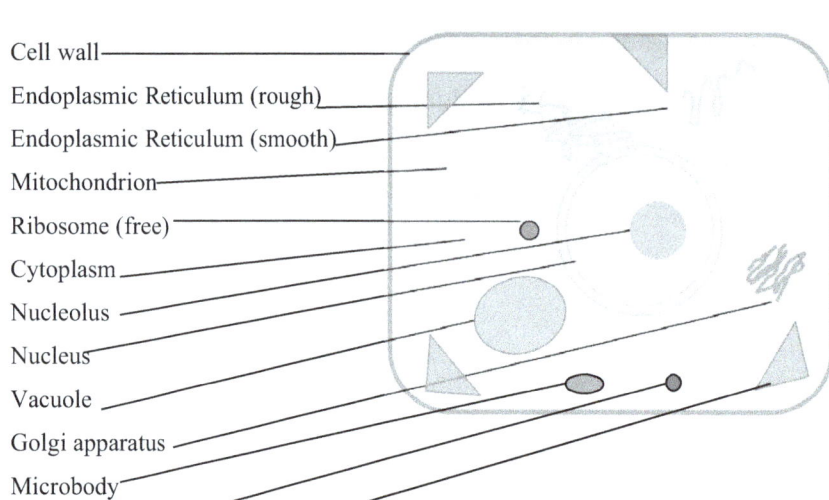

Generalized Eukaryotic (Plant) Cell

COMPARISON OF PROKARYOTIC AND EUKARYOTIC CELLS

Prokaryotic cells are the simplest organisms. They are small and encased by a cell membrane and a rigid cell wall. Since they lack membrane-bounded organelles, all cytoplasmic constituents have access to all areas of the cell. There is usually no true nucleus. The plasma membrane carries out some of the functions organelles perform in eukaryotic cells. Some prokaryotes are propelled by a rotating flagellum. Prokaryotes cause disease, and are involved in many industrial processes. There are two main groups of prokaryotes: archaebacteria and bacteria.

Eukaryotic cells have no cell wall; however, they have an extensive endomembrane system that creates a highly compartmentalized cell with numerous membrane-bounded organelles that carry out specialized functions. Plant cells normally contain a large central vacuole (for storage) and thick cell walls. Animal cells often secrete an extracellular matrix of glycoproteins to help coordinate the behavior of all the cells in a tissue.

PROPERTIES OF CELL MEMBRANES

Basically, all cells have a cell membrane. The plasma membrane, composed of a **phospholipid bilayer**, encloses a cell and separates its contents from its surroundings. The phospholipid bilayer impedes the passage of any water-soluble substances through it since the nonpolar tails are oriented away from water and the polar heads are oriented toward the water causing hydrogen bonding to hold the membrane together. Transport proteins in the plasma membrane help molecules and ions move into and out of the cell. The cell membrane also contains receptor proteins which induce changes in the cell when the cell comes in contact with specific molecules and markers that identify the cell. Passive transport across membranes moves down the concentration gradient. Passive transport processes include diffusion to move oxygen into cells, facilitated diffusion to move glucose into cells, and osmosis of water into cells placed in a hypotonic solution. Bulk transport utilizes endocytosis and exocytosis. Active transport, the movement of a solute across the membrane against the concentration gradient, requires the use of highly selective protein carriers and the expenditure of ATP for energy.

Structure and function of cell organelles in cells and whether they are found in Prokaryotic (Pro), or Eukaryotic [Animal (An), and Plant (Plt)] cells.

Structure	Function	Pro	An	Plt
Cell wall	Protection; support	Yes	No	Yes
Cytoskeleton	Structural support; cell movement		Yes	
Flagella (cilia)	Motility or moving fluids over surfaces	Maybe	Maybe	Mostly absent
Plasma Membrane	Regulates what passes into and out of cell; cell to cell recognition	Yes	Yes	Yes
Endoplasmic Reticulum (ER)	Forms compartments and vesicles; participates in protein and lipid synthesis	No	Yes	Yes
Centrioles	Help assemble microtubules	No	Yes	No
Nucleus	Control center of cell; directs protein synthesis and cell reproduction	No	Yes	Yes
Golgi Apparatus	Packages proteins for export from cell; forms secretory vesicles	No	Yes	Yes
Lysosomes	Digest worn-out organelles and cell debris; play role in cell death	No	Yes	Yes
Microbodies	Isolate particular chemical activities from rest of cell	No	Yes	Yes
Mitochondria	Site of oxidative metabolism	No	Yes	Yes
Chloroplasts	Site of photosynthesis	No	No	Yes
Chromosomes	Contain hereditary information	Single circle	Many	Many
Nucleolus	Assemble ribosomes	No	Yes	Maybe
Ribosomes	Site of protein synthesis	Yes	Yes	Yes
Vacuoles	Storage compartment for water, sugars, ions, pigments	No	No	Yes

There can be anywhere from thousands to millions of ribosomes in any given cells. Ribosomes are responsible for protein synthesis, or in other words, they build proteins for the cell to use. There are two parts to a ribosome, a large subunit and a small subunit. The two parts fit together to make a complete structure and use mRNA and rRNA to complete their task. The mRNA, or messenger RNA, threads between the two parts of the ribosome and contains instructions from the DNA on how to create the needed protein. The rRNA, also called ribosomal or transfer RNA, collects the necessary amino acids.

 Enzymes

ENZYME-SUBSTRATE COMPLEX

Enzymes act as catalysts in biological processes. Most enzymes are globular proteins with one or more three-dimensional active sites where a specific substrate can bind, forming an enzyme-substrate complex. The 3D shape of the enzyme enables it to stabilize a temporary association between substrates. Either by bringing two substrates together in a correct orientation or by stressing particular chemical bonds of a substrate, the enzyme lowers the activation energy needed for the new bonds to form. Once the bonds of the substrates are broken, or new bonds are formed, the substrates convert to products. These products then dissociate from the enzyme. The reaction can then proceed much more quickly and with much less energy than would have been needed without the enzyme.

Example:

>Step 1: The substrate, sucrose, consists of glucose and fructose bonded together and the enzyme has a very specifically-shaped active site.

>Step 2: The substrate binds to the enzyme to form an enzyme-substrate complex.

>Step 3: Stress is placed on the glucose-fructose bond in sucrose; the bond breaks and the products, glucose and fructose are released. The enzyme is unaffected and becomes free to bind to another sucrose.

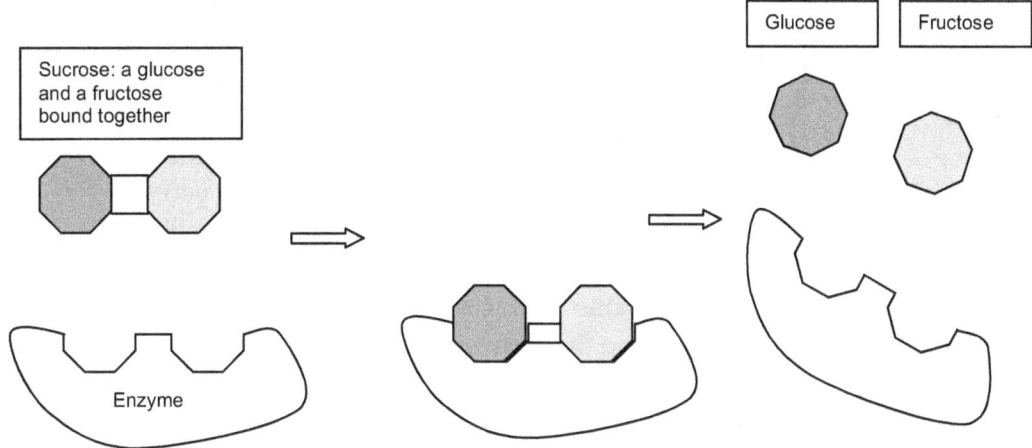

Factors affecting enzyme activity are concentration of the enzyme, concentration of the substrate, temperature, pH, salt concentration and the binding of specific regulatory molecules. Many reactions are multi-step so are better facilitated using multienzyme complexes that are noncovalently bonded assemblies of enzymes that catalyze different steps of the sequence. Not all biological catalysts are proteins; RNA itself can act as a catalyst in certain types of reactions.

INORGANIC COFACTORS

The activity of enzymes is often facilitated by cofactors which can be metal ions or other substances. The active sites of many enzymes contain metal ions such as zinc, molybdenum and manganese to help draw electrons away from substrate molecules. This makes the bonds less stable and easier to break. These metals and many vitamins are required in the diet so that they are present to play their roles.

ROLES OF COENZYMES

Cofactors that are non-protein organic molecules are called coenzymes. Coenzymes often shuttle hydrogen ions (H^+) from one enzyme to another in a cell. In oxidation-reduction reactions catalyzed by enzymes, a pair of electrons passes from the enzyme to the coenzyme that becomes the electron acceptor. The coenzyme passes the electrons to a different enzyme which releases them (plus energy) to a substrate in another reaction.

A good example of this is the coenzyme nicotinamide adenine dinucleotide (NAD^+). The NAD^+ is composed of NMP (nicotinamide monophosphate) and AMP (adenine monophosphate) joined phosphate-to-phosphate. AMP acts as the core and provides a shape recognized by many enzymes. NMP contributes a site that accepts electrons (is reduced). $NAD^+ + 2$ electrons $+ 1$ proton → NADH

The energy of NADH is transferred to other molecules when the C-H bonds are broken.

INHIBITION AND REGULATION

A substance that binds to an enzyme and decreases its activity is called an inhibitor. If the end product of a biochemical pathway acts as an inhibitor of an earlier reaction in the pathway, that product is a feedback inhibitor. Competitive inhibitors compete with the substrate for the same active site, and noncompetitive inhibitors bind to the enzyme in a location other than the active site. Competitive inhibitors displace substrate molecules from the enzyme by already being bound to the active site, while noncompetitive inhibitors cause the enzyme to change shape which changes the shape of the active site. Many noncompetitive inhibitors bind to the enzyme location called an allosteric site which serves as a chemical active/inactive switch. Allosteric inhibitors bind to an allosteric site and reduce the enzyme activity while activators bind to an allosteric site and increase the enzyme activity.

Biochemical pathways are a series of steps which are coordinated and regulated by a cell. The product of one step becomes the substrate for the next step. If an excess of one of the products is made, it may bind to an allosteric site on an enzyme that catalyzes an earlier reaction to slow (or stop) the reaction, causing feedback inhibition.

$$\begin{array}{c} E_1 \; E_2 \; E_3 \; E_4 \; E_5 \\ A \to B \to C \to D \to E \to P \end{array}$$

Energy Transformations

THE AEROBIC PATHWAY

In aerobic (with oxygen) respiration, the cell gains energy from glucose molecules in a sequence of four stages: glycolysis, pyruvate oxidation, the Krebs cycle and the electron transport chain. Oxygen is the final electron acceptor.

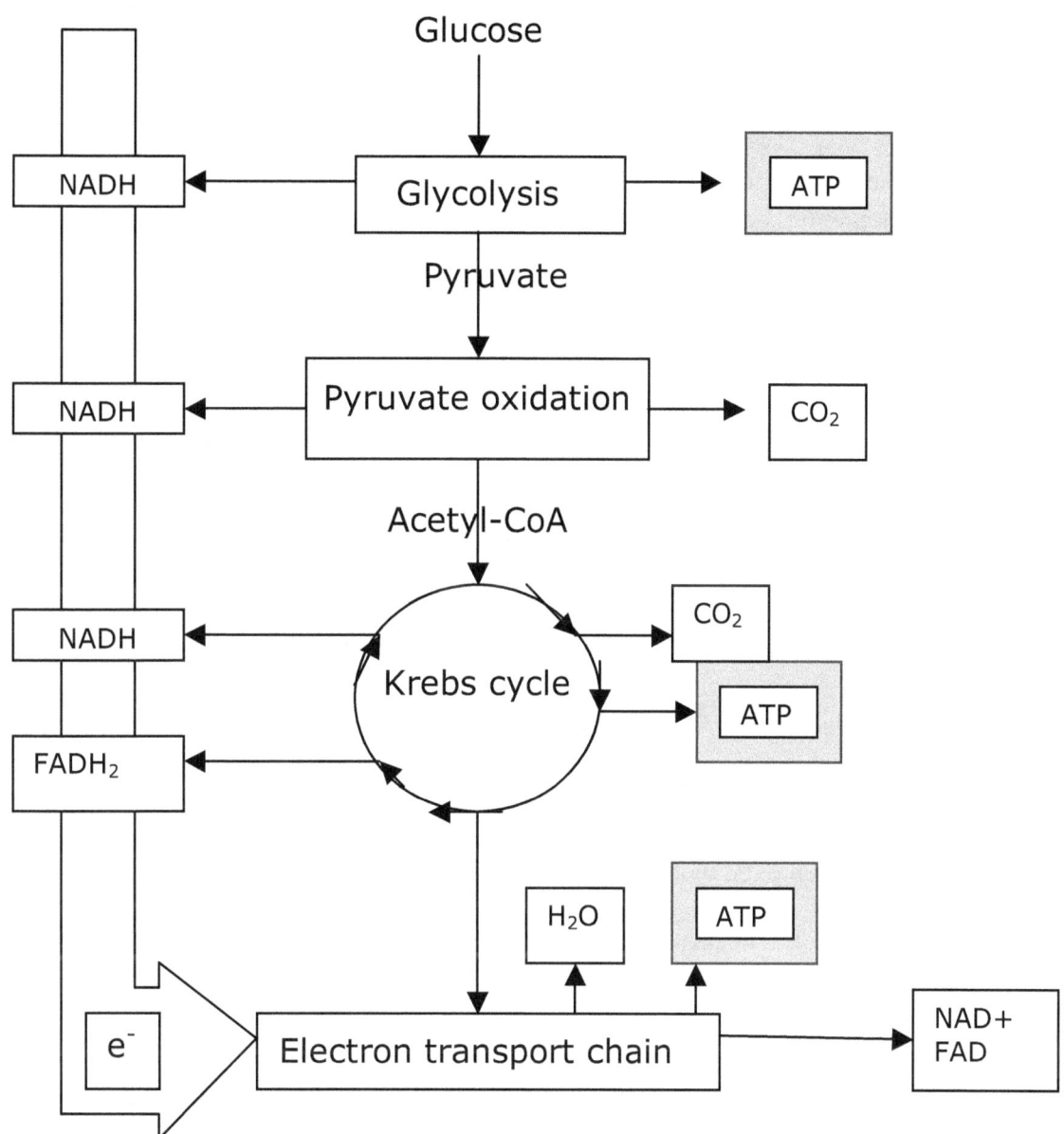

Aerobic prokaryotes and the mitochondria of eukaryotes produce the vast majority of their ATP energy by aerobic respiration, in which a proton gradient formed from electrons harvested from organic molecules are donated to oxygen.

GLYCOLYSIS AND ANAEROBIC PATHWAYS

Stage one of aerobic respiration and the only stage of anaerobic respiration is glycolysis, a 10-reaction biochemical pathway. It uses substrate-level phosphorylation in the cytoplasm of the cell. In the process, two ATP's are used up, but four ATP's are produced for each glucose molecule that is catabolized. Thus, the net oxidation (catabolism) of glucose is represented as:

$$C_6H_{12}O_6 + 6\ O_2 \rightarrow 2\text{ pyruvate molecules} + 6\ H_2O + 2\text{ ATP (energy)}$$

In the absence of oxygen, anaerobic respiration continues into lactic acid or ethanol fermentation. Examples of ethanol fermentation are yeast bread, beer and wine. Lactic acid fermentation takes place in muscles and contributes to muscle fatigue when the resulting lactate cannot be removed fast enough by the blood.

AEROBIC RESPIRATION

Aerobic respiration involves pyruvate oxidation (stage two), the Krebs cycle (stage three), and the electron transport chain (stage four). Pyruvate, the end product of glycolysis, passes through the outer mitochondrial membrane where CO_2 is split off of the pyruvate to make an acetyl group and Vitamin C is turned into Coenzyme-A. The acetyl group and Coenzyme-A combine to form acetyl-CoA, releasing NADH and a proton (H^+).

$$\text{Pyruvate} + NAD^+ + CoA \rightarrow \text{acetyl-CoA} + NADH + H^+ + CO_2$$

One molecule of NAD^+ is reduced to NADH to carry electrons to be used to make ATP.

Acetyl-CoA feeds into a multi-step-reaction cycle in the Krebs cycle (citric acid cycle) where two more ATPs are extracted by phosphorylation, and a large number of electrons are removed by reduction of NAD^+ to NADH.

$$\text{Acetyl-CoA} + \text{oxaloacetate} + FAD + ADP + 3\ NAD^+ \rightarrow$$
$$2\ CO_2 + 3\ NADH_2^+ + ATP + FADH_2$$

Remember that each glucose yields two pyruvates, thus two Acetyl-CoA molecules, so the resultant energy per glucose molecule at the end of the Krebs cycle is 6 $NADH_2^+$ (12 electrons), 2 $FADH_2$ and 2 ATP.

In the electron transport chain, $NADH_2^+$ starts a chain of electron transfers that end up producing ATPs. A series of steps in the inner membrane of the mitochondria release energy to make O_2 and to pump protons across the membrane. The resulting proton gradient is used by ATP synthase and ADP to produce ATP. The end result of aerobic respiration is 36 ATPs; 34 more than glycolysis alone (anaerobic respiration).

$$C_6H_{12}O_6 + 6\ O_2 \rightarrow 2\ CO_2 + 6\ H_2O + 36\text{ ATP (energy)}$$

The relative levels of ADP and ATP regulate the catabolism of glucose at a key reaction of glycolysis and a key reaction of the Krebs cycle using feedback inhibition.

The electron transport chain is located in the mitochondria. It is a series of carrier atoms which pass along electrons. At each stage, energy is released and used in ATP production. It begins with NADH and FADH2 which deliver the electrons. After passing through other carriers, the electrons reach the final carrier in the chain, which is oxygen. The oxygen combines with hydrogen, resulting in water.

PHOTOSYNTHESIS

There are two phases of photosynthesis. One is the light phase which uses the chlorophyll in the chloroplast of a plant cell. The chloroplast gives plants their green color. Cells containing chlorophyll use photosynthesis to make food out of nonliving materials which serves as the support for all animal life. The light phase changes sunlight energy into chemical energy in the form of ATP and NADPH, which are used to split water in hydrogen and oxygen. The oxygen is set free and becomes the oxygen all animals need to breathe. The hydrogen is retained and used in the dark phase to combine with carbon dioxide in a process known as carbon fixation to form a simple sugar like glucose. By-products include starch, plant oils and proteins. The summary equation for photosynthesis is:

$$6\ CO_2 + 6\ H_2O + light + chlorophyll \rightarrow C_6H_{12}O_6\ (glucose) + O_2$$

Anaerobic photosynthesis evolved in the absence of oxygen. Instead of reshuffling chemical bonds (in glycolysis), some organisms use light to pump protons out of their cells and use the resulting proton gradient to power the production of ATP through chemiosmosis. These organisms use H_2S present in the oceans as a source of hydrogen atoms (see anaerobic respiration).

Cell Division

STRUCTURE OF CHROMOSOMES

The nucleus contains various bodies called chromosomes. They may be threadlike, rodlike, or have a number of other shapes. They contain chromatin, a complex of 40% DNA and 60% protein, which contains the determiners of heredity (genes). A gene is a small mass of nucleoprotein. Chromosomes differ widely among species and even among individuals of the same species. The particular array of all the chromosomes an individual possesses is called its karyotype. By definition, the number of chromosomes in a species is the number of one complete set of chromosomes, the haploid (*n*) number. The normal number of chromosomes in a cell of a given species is the diploid (*2n*) number. An animal usually receives one paternal and one maternal chromosome.

THE CELL CYCLE

The cell cycle consists of five phases: G_1, the primary growth phase; S, the replication phase; G_2, the second growth phase; mitosis (comprised of prophase, metaphase, anaphase and telophase); and cytokinesis. Interphase is the time between cell divisions. It includes the G_1 phase, which occurs after cell division and before DNA replication and occupies more than half the cell cycle. The S phase, also part of interphase, is when the cell's DNA is copied. And the G_2 phase, the third part of interphase, occurs after the DNA is synthesized and before cell division. At any given time, the majority of cells in an animal's body are in the G_0 phase which is a resting state.

MITOSIS

Mitosis is the division of the nucleus during the four-phase cell division to form two identical daughter cells. When accompanied by cytokinesis, the two identical daughter cells are fully formed. The four phases of mitosis are prophase, metaphase, anaphase and telophase.

At the start of prophase, there is a shortening and tight coiling of the DNA into rod-shaped chromosomes. The two copies of each chromosome (called chromatids) stay connected to each other by the centromere. The nucleus and nucleolus membranes break down and disappear. Centrosomes (two pairs of dark spots) appear next to the disappearing nucleus. In animal cells, each centrosome contains a pair of small cylindrical bodies called centrioles (not found in plant cells). As centrosomes separate and move toward opposite poles of the cell, spindle fibers made of microtubules radiate from the centrosomes (now called mitotic spindle) in preparation for mitosis. Kinetochore fibers are attached to a disk-shaped protein called a kinetochore that is found in the centromere region of each chromosome and extend from the kinetochore of each chromatid to one of the centrosomes. Polar fibers extend across the dividing cell from centrosome to centrosome.

During <u>metaphase</u>, the chromosomes are easier to identify; karyotypes are typically made from photomicrographs of chromosomes in metaphase. Kinetochore fibers move the chromosomes to the center of the dividing cell and hold them in place there along a perceived metaphase plate at the equator of the cell.

The <u>anaphase</u> stage is the shortest, but most beautiful. It starts when all the chromosomes divide simultaneously. Then the two chromatids of each chromosome separate at the centromere and slowly move, centromere first, toward the opposite poles where their kinetochores are attached. Two forms of movement take place simultaneously, each driven by microtubules: the poles move apart and the centromeres move toward the poles as the microtubules that connect them shorten. The microtubules are progressively disassembled, pulling the chromatids even closer to the poles of the cell. When the sister chromatids separate in anaphase, the essential element of mitosis – accurate partitioning of the replicated genome is complete.

In <u>telophase</u>, after the chromosomes reach opposite ends of the cell, the spindle fibers disassemble. The microtubules break down to be re-used to construct the cytoskeletons of the daughter cells. A nuclear envelope forms around each set of chromosomes and the chromosomes begin to uncoil to permit gene expression. An early group of genes to be expressed is the rRNA, prompting the nucleolus to reappear. Mitosis is complete at the end of telophase.

CYTOKINESIS IN PLANTS AND ANIMALS

The cytoplasm of the cell divides in a process called cytokinesis. While mitosis was in progress, the cytoplasmic organelles were realigned to give at least one to each daughter cell. In animal cells, cytokinesis begins with the microfilaments pinching the cell membrane inward midway between the dividing cell's two poles (cleavage furrow). In plant cells, the vesicles formed by the Golgi apparatus fuse at the midline of the dividing cell, forming a membrane-bound cell wall called the cell plate.

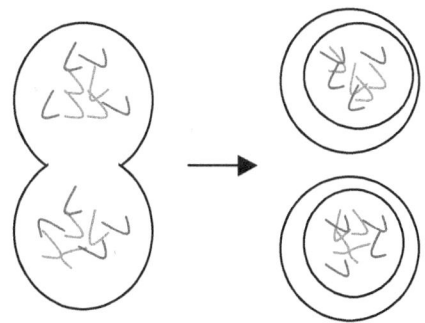

The two offspring cells are approximately the same size. Each receives an identical copy of the original cell's chromosomes and approximately half of the original cell's cytoplasm and organelles.

MEIOSIS

Meiosis is the process of nuclear division that reduces the number of chromosomes in new cells to half the number in the original cell. In humans, meiosis produces haploid reproductive cells called gametes. Cells undergo the G_1, S and G_2 phases of interphase so that the cell grows to a mature size and copies its DNA. Cells undergoing meiosis divide twice so one diploid cell ($2n$) becomes 4 haploid cells ($1n$). Meiosis in diploid organisms has Meiosis I and Meiosis II, each with prophase, metaphase, anaphase, telophase and cytokinesis. The purpose of Meiosis I is to produce two cells, each of which has half the number of chromosomes (but each chromosome has two chromatids). Meiosis II has no copying of the DNA. The purpose of Meiosis II is to create a total of four cells, each with a single chromatid of each chromosome from Meiosis I.

In Prophase I, the DNA tightly coils into chromosomes, spindle fibers appear, and the nucleus and nucleolus disassemble. The two copies of each chromosome (called chromatids) stay connected to each other by the centromere. Chromosomes line up next to their homologues (called synapsis) forming a tetrad. During synapsis, the chromatids within a homologous pair twist around one another (forming a chiasma), allowing portions of chromatids to break off and attach to adjacent chromatids on the homologous chromosome in a process called crossing-over. This exchange of genetic material between maternal and paternal chromosomes results in genetic recombination.

During Metaphase I, tetrads line up randomly along the midline of the dividing cell, the metaphase plate. Spindle fibers from one pole attach to the centromere of one homologous chromosome, and spindle fibers from the other pole attach to the centromere of the other homologous chromosome of the pair. This looks similar to metaphase in mitosis, except that the spindle microtubules are only able to attach to kinetochore proteins on the outside of each centromere.

In Anaphase I, each homologous chromosome (consisting of two chromatids attached by a centromere) moves to an opposite pole of the dividing cell. Random separation of the homologous chromosomes is called independent assortment.

As Telophase I and Cytokinesis I progress, chromosomes reach the opposite ends of the cell and form a cluster at each pole. The nuclear membrane re-forms and the parent cell divides the organelles and cytoplasm between the two daughter cells. New cells contain a haploid number of chromosomes but two copies of each chromatid.

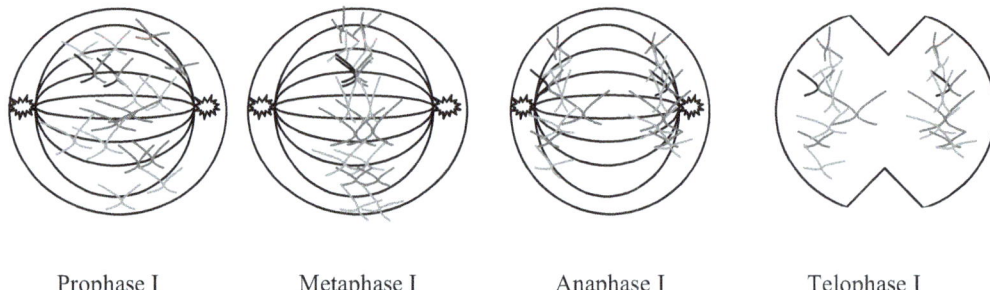

Prophase I   Metaphase I   Anaphase I   Telophase I

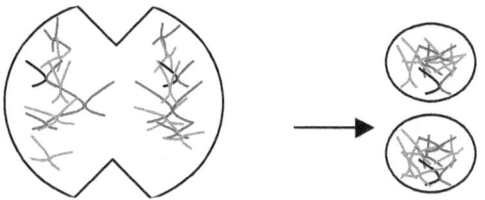

Cytokinesis I

After a brief interphase with no replication of DNA, prophase II starts. The spindle fibers form and begin to move the chromosomes toward the midline of the dividing cells.

In metaphase II, the chromosomes (consisting of sister chromatids) move to the midline of the dividing cell, facing opposite poles of the dividing cell. Kinetochore microtubules from opposite poles attach to opposite sides of the same centromere.

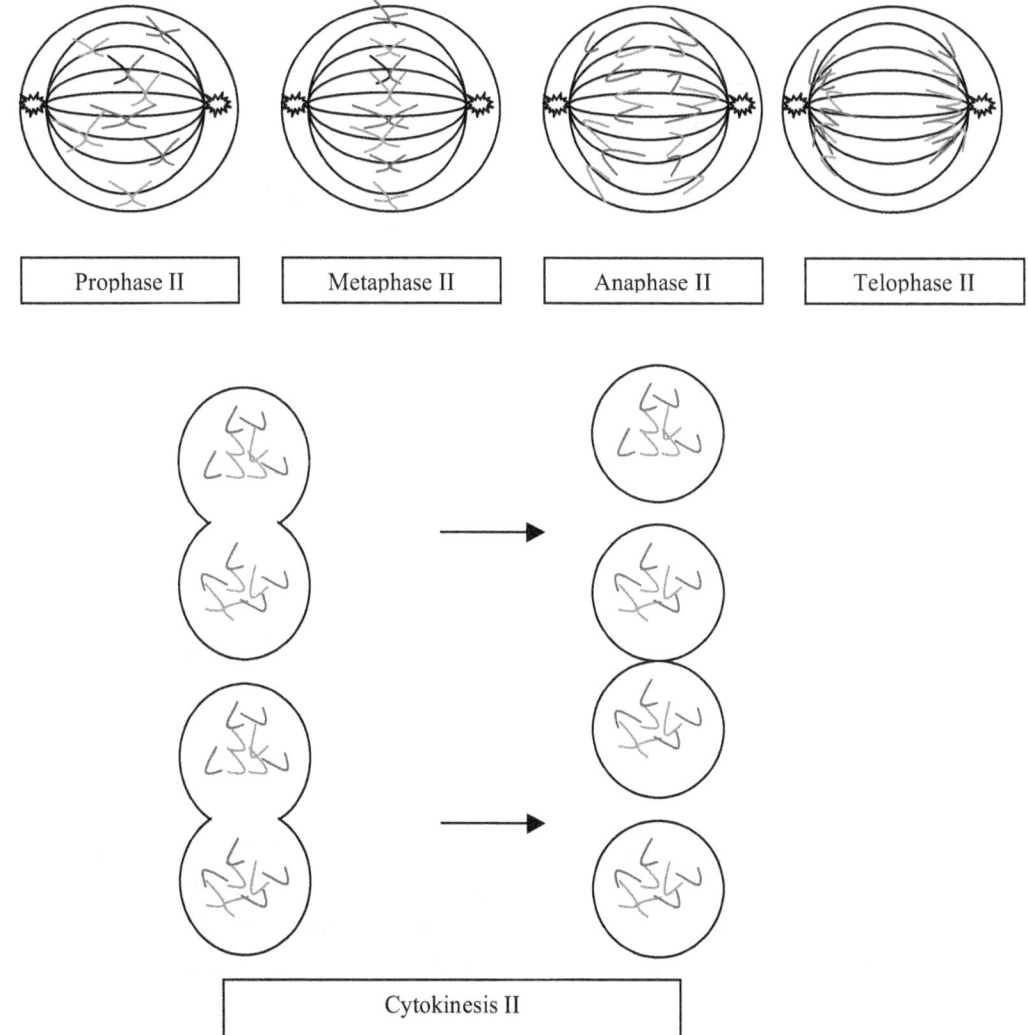

| Prophase II | Metaphase II | Anaphase II | Telophase II |

Cytokinesis II

In anaphase II, the chromatids separate and move toward opposite poles of the cell.

During telophase II and cytokinesis II, the nuclear membrane forms around the cluster of chromosomes in each of the four new cells. After cytokinesis II, there are four new cells each containing half of the original cell's number of chromosomes. No two cells are alike due to genetic recombination.

In animals, meiosis produces haploid reproductive cells called gametes. Meiosis only occurs in reproductive organs (testes in males and ovaries in female). Spermatogenesis is the production of four male gametes known as sperm cells in the testes. Each diploid reproductive cell divides meiotically to form four haploid cells called spermatids, each of which then develops into a mature sperm cell. Oogenesis is the name of the female's production of one mature egg cell or ovum. A diploid reproductive cell divides meiotically to produce one mature egg cell (ovum) and three polar bodies which degenerate (die). During cytokinesis I and cytokinesis II of oogenesis, the cytoplasm of the original cell is divided unequally between new cells, giving the ovum the majority.

Asexual reproduction is the production of offspring from one parent rather than from two parents and does not usually involve meiosis or the union of gametes. In unicellular organisms, such as bacteria, new organisms are created by either binary fission or mitosis. In multicellular organisms, asexual reproduction results from the budding off of portions of their bodies so that each "offspring" is genetically identical to the one parent.

Sexual reproduction is the production of offspring through meiosis and the union of a sperm and an egg. Offspring produced by sexual reproduction are genetically different from the parents because genes are combined in new ways in meiosis. Except for identical twins, sexually produced offspring contain unique combinations of their parents' genes. This enables a species to adapt rapidly to new conditions.

 *Chemical Nature of the Gene*

WATSON-CRICK MODEL OF NUCLEIC ACIDS

In 1953, James Watson and Francis Crick (Cambridge University) proposed that the structure of the DNA molecule was two chains of nucleotides that are intertwined to make a double helix. Each strand is made up of repeating sugar and phosphate units joined by phosphodiester bonds. The two strands are wrapped around a common axis similar to two strands of rope, held together with crossbars.

A nucleotide is composed of a phosphate group, a deoxyribose group (5-carbon sugar) and a nitrogen-containing base (either purine or pyrimidine). A DNA molecule can have as many as 200,000 nucleotides. The two strands are held together by weak hydrogen bonds between a purine and its opposing pyrimidine.

Because of the shapes of the purines and pyrimidines, and the fact that hydrogen bonds function only over short distances, adenine can pair only with thymine and guanine will pair only with cytosine. Therefore, the number of thymines in a particular kind of DNA will equal the number of adenines, and likewise for guanines and cytosines. The nucleotides of a single strand may be linked together in any arrangement, and each nucleotide can be repeated as often as desired. Once the order of one strand is established, it determines the order of the other strand. The total amount of DNA in similar kinds of cells remains constant from generation to generation, implying that both the quality and the quantity of the DNA remains the same in similar cells derived from the same parent cell.

DNA REPLICATION

The Watson-Crick model suggests that the copying of DNA is complementary since the order of one strand determines the order of the other strand. Replication is accomplished by "unzipping" the two strands of the helix. Each strand can then add nucleotides to itself in the correct complementary order, using various enzymes. After replication, the two identical DNA's each contain one strand that is original and one that is new. The stages of replication are initiation, elongation and termination.

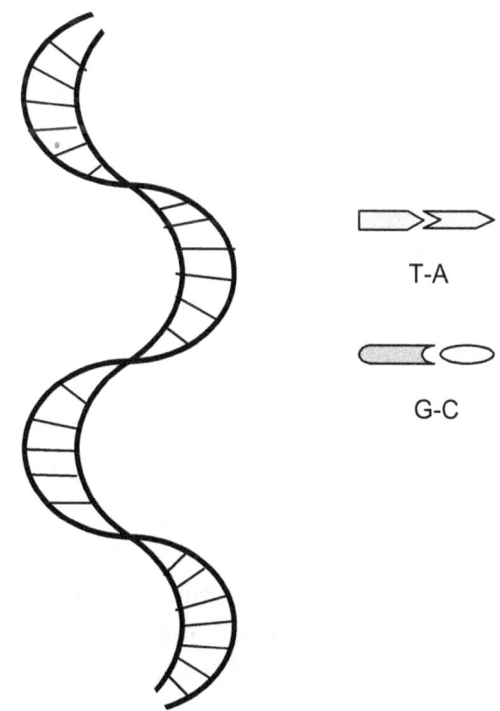

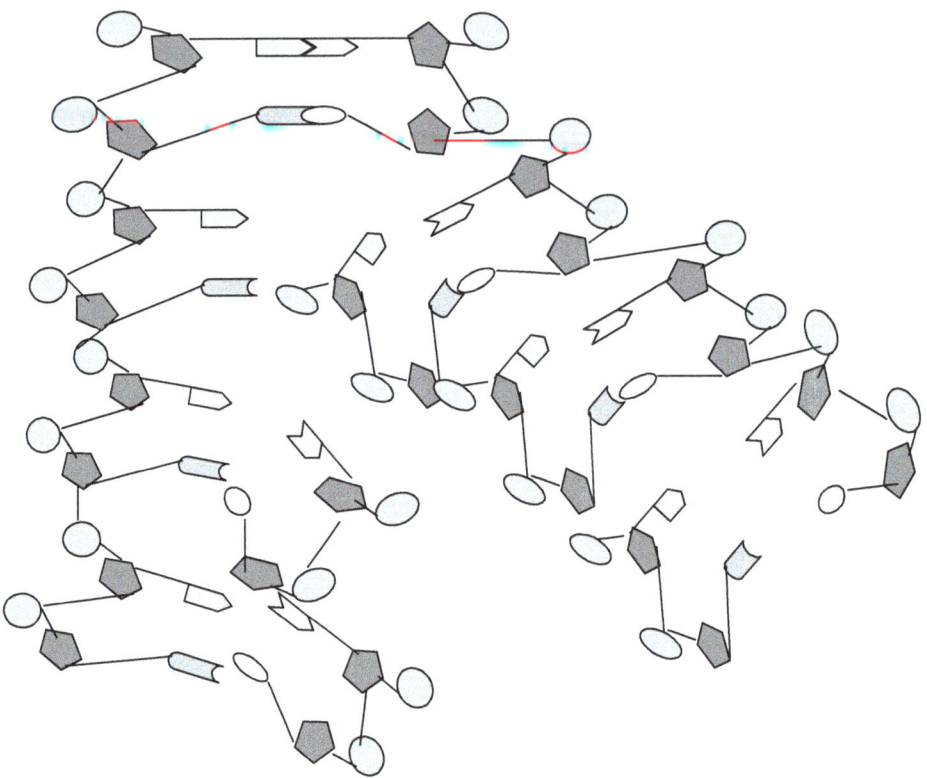

MUTATIONS

A mutation is a permanent change in a cell's DNA. It can be a change in nucleotide sequence, alteration of gene position, gene loss or duplication, or an insertion of one or several foreign sequence(s). One type of mutation at the chromosomal level is an inversion of genes where a specific section is reversed. Another type is translocation of genes, where information from one of two homologous chromosomes breaks and binds to the other. This type is usually lethal. Alterations of DNA sequence include deletion, insertion, inversion and substitution. Negatively, mutations are the cause of many defects and diseases and can cause death. Positively, mutations increase genetic diversity and allow for a wider variation of traits for natural selection.

CONTROL OF PROTEIN SYNTHESIS: TRANSCRIPTION, TRANSLATION, POSTTRANSCRIPTIONAL PROCESSING

The information in genes is expressed in two steps. First, it is transcribed into RNA using tRNA and rRNA. Ribosomal RNA (rRNA) provides the site during protein synthesis where polypeptides are assembled. Transfer RNA (tRNA) transports amino acids to the ribosome to be used in building the polypeptides and position each amino acid at the correct place. Messenger RNA (mRNA) molecules are long strands of RNA that are transcribed from DNA and travel to the ribosomes to direct precisely which amino acids are assembled into polypeptides. Each "word" or codon of three nucleotides in a specific sequence corresponds to one of the twenty amino acids.

STRUCTURAL AND REGULATORY GENES

DNA, the molecule responsible for the inheritance of traits, is divided into functional units called genes. Gene expression is controlled by regulating transcription either at the transcriptional or posttranscriptional levels. The most common method of control is at the transcriptional level. Regulatory proteins identify DNA sequences without unwinding the helix by inserting DNA-binding motifs into the major groove where the edges of the base-pairs are exposed.

In prokaryotic gene regulation, repressors are proteins that bind to regulatory sites on DNA and prevent or decrease initiation of transcription, and activators are proteins that bind to regulatory sites on the DNA to stimulate the initiation of transcription.

Eukaryotic transcription factors include basal transcriptional factors and specific transcriptional factors, enhancers, coactivators and mediators. Posttranscriptional control of gene expression is exercised by proteins and small RNAs. Proteins interacting with small RNA can carry out alternative splicing of RNA transcripts. Chemical alterations of specific bases cause TNA transcript editing. Translation repression and selective degradation of mRNA transcripts provide further control of gene expression.

TRANSFORMATION

Plant transformation is the stable introduction of a gene into a plant genome using a multifaceted protocol that requires a gene-delivery system, and a reliable culturing system for regenerating plant tissue sections into mature plants.

Gene transformation is the process of introducing genes into plants by methods which by-pass the sexual seed production process. Essentially, it is a process by which genes are "cut" from the cells of one organism and "pasted" and integrated into the cells of another organism. Once the cells are transformed, they are grown into new plants capable of "expressing" a desired characteristic. Genes that are resistant to specific diseases, antifungal, antibacterial or resistant to toxicity are the most desirable to use for transformation.

VIRUSES

The viruses of eukaryotes are similar to prokaryote-infecting viruses. Proviruses are viral DNA integrated into the host cell. Some of the DNA viruses can either initiate an infection (lytic in prokaryotes) cycle or can form proviruses. Simian Virus 40 (SV40) causes cancer in hamsters but not in its normal hosts. SV40 and a number of other viruses can introduce new functional genes into the host DNA.

Retroviruses can also insert their nucleic acid into host DNA via reverse transcriptase which is carried with the RNA into the host cell. The inserted viral DNA makes RNA transcripts which are packaged with viral protein coats and reverse transcriptase. The viral DNA may, depending on its insertion point, cause mutations of the host DNA. Most viral DNA insertions do not damage the host, but rather become part of the host genome and can be passed on if they have managed to infect a germ-line cell.

Cancer is a disease in which cells escape the restraints on normal cell growth. Cancer is an inheritable disease (at least from cell to daughter cells). Once a cell has become cancerous, all of its descendant cells are cancerous. Gross chromosomal abnormalities are often visible in cancerous cells. Most carcinogens (cancer-generating factors) are also mutagens (mutation-generating factors). Oncogenes are genes resembling normal genes but in which something has gone wrong, resulting in a cancer.

Viruses seem able to cause cancer in three ways. Presence of the viral DNA may disrupt normal host DNA functions. Viral proteins needed for virus replication may also affect normal host gene regulation. Since most cancer-causing viruses are retroviruses, the virus may serve as a vector for oncogene insertion. Viruses can thus serve as a possible vector to place healthy (non-mutated) alleles into eggs.

Photosynthesis

Photosynthesis is how plants take the energy from the sun and turn it into food for the plant. Energy starts with the sun that the plants absorb through the leaves. The chlorophyll in the leaf absorbs the sunlight and changes the water and carbon dioxide into oxygen and glucose. When the plants create sugars and starches, this process combines atoms into a large atom. On the bottom of the leaves are the stomata, which are small openings. The **stomata are where transpiration takes place.** The stomata open up to let air in and release oxygen. They also let water out. Transpiration is the evaporation of water from plants when the stomata are open during photosynthesis. Photosynthesis is the process of a plant taking in carbon dioxide via the stomata and creates food for the plant, and oxygen for the planet. Plants need water (H_2O) and carbon dioxide (CO_2) to make food via photosynthesis.

Food Chain

Plants are called **primary producers**. **Herbivores** are animals that eats the plants. They are called **primary consumers** because they are what eat the plant. **Carnivores** are meat eaters and are **secondary consumers** because they eat a primary consumer. A **tertiary** consumer is a carnivore that eats another carnivore. The chemical process from eating food results in free energy. Eating the plant breaks the energy bond.

Although food chains and food webs depict the flow of energy through an ecosystem, they do not factor in one crucial element. This is the concept of biomass.

Biomass is determined by multiplying the number of animals by their weight. This is important when considering that a food web shows that a bird consumes a worm. This is true, a bird will eat a worm, but a bird will need more than just one worm to survive. This means that the biomass of worms must be greater than the biomass of birds in an ecosystem. It also means that the largest amount of stored energy is contained at the lowest levels of the food chain.

This concept is often organized into a biomass pyramid. Producers are the lowest level of the pyramid, with the greatest amount of biomass and energy. Generally herbivores follow producers, then carnivores. If the pyramid becomes off balance, then the ecosystem will not be able to support all of the organisms.

Biology Classifications and Definitions

Taxonomy: the science of naming organisms

Monera: a single cell bacteria

Protista: multi cell organisms like amoebae

Fungi: mushrooms

Plantae: plants

Animalia: animals

There are more insects than any other animal.

This is the scientific classification order for organisms:
1. Kingdom
2. Phylum (animals) or Division (plants)
3. Class
4. Order
5. Family
6. Genus
7. Species

For Humans, the complete classification is as follows in **bold:**
1. Kingdom **Animalia**
2. Phylum (animals) or Division (plants) **Chordata**
3. Class **Mammalia**
4. Order **Primates**
5. Family **Hominidae**
6. Genus **Homo**
7. Species **Sapiens**

Animals with a spinal cord are called **Chordata**. A **phylogenic tree** is something that predicts probable evolution.

Homologous structures: similar bone structures. A human arm is similar to other mammal's arms, like a monkey's.

Ecosystem

Ecosystem is the most inclusive ecological term and includes air, water and energy. It sustains life. Without energy all life will cease.

The sun produces all (most) energy. 1.2% of energy from the sun is used in photosynthesis. It takes 809 grams of plants per square meter to support 1.5 carnivores. Developing countries could benefit by being all vegetarian as it takes much less natural resources for people to eat plants versus eating meat, which takes more resources to create.

Chemical energy is a secondary form. Covalent bonds are caused by atoms sharing electrons.

O-C=O (each line for each electron which is shared). Not all covalent bonds are the same.

Decomposer is a fungus or bacteria that absorbs nutrients. It then respires (excretes).

Oxygen processes food, excretes H20 and carbon dioxide. Plants breathe carbon dioxide.

A **biome** is a certain type of ecosystem based on several factors including humidity, heat and vegetation. Some of the major biomes are tropical rain forest, tundra, taiga, temperate grassland, temperate forest, desert, tropical rain forest, tropical dry forest and savanna. In the desert you experience the most rapid temperature changes. The night is very cold as it has no way to hold the heat in, while during the day, the air is stifling. In the tropical rainforest, farmers are clearing the tropical rainforest and are growing grassland for cattle. This is because they make more money from selling the cattle, but it has major effects on the environment.

Charles Darwin

Charles Darwin sailed a ship around the world called the HMS Beagle. He found fossils of extinct organisms. He also went to the Galapagos Islands. Darwin studied finches, and he discovered that they had different beak sizes. It was noticing the beaks that helped Darwin realize natural selection. Bigger beaks helped them to survive and so he learned of evolution and natural selection. Natural selection is when nature determines that a species will not survive. For example, an animal that is not fast enough to run from a predator will not survive to procreate, so eventually (over hundreds or thousands of years) the animals left to procreate are the fastest and have inherited that ability from their parents.

Genetics

Gregor Mendel was the father of genetics and inheritance. He was a priest in charge of a garden who became interested in some of the plants. Pea pods are self-fertilizing. Three of four seeds were purple, and one seed out of four was white. Each plant contained two genetic codes. The purple was dominant over white because it showed up more often. Look at the chart below to determine the chances of getting each kind of plant:

PP=Purple (Upper case means dominant, lower case means recessive)
pp=White

	Purple	White
Purple	PP	P*p*
White	*p*P	*pp*

The inheritance of sex is determined by the x and the y chromosome. All chromosomes come in pairs. For use in the chart, females are XX and males are XY.

	X	X
X	XX	XX
Y	YX	YX

To get the values (YX), all you do is add the intersection of each row and column on the table. According to the table on the previous page (and biology), the chances of having either a boy or girl are 50%.

In order to understand a Punnett square (the tables above and below), it is important to first understand the significance of phenotypes and genotypes. A phenotype is the physical appearance of the gene. For example, eye color, hair color and earlobe type are all phenotypes. Phenotypes are determined by genotypes, which are the basically the genetic appearance of the trait, or in other words, what alleles are present. Genotypes are expressed as a combination of two alleles. When making Punnett squares, a trait is assigned a letter. The alleles are represented by capital or lowercase letters. An allele represented by a capital letter represents a dominant trait, while an allele represented with a lowercase letter represents a recessive trait. If a trait is dominant, it means that it exerts phenotypic control and masks the recessive trait.

A Punnett square is a diagram which displays all of the possible crosses between an egg and sperm cell. Punnett squares are used to show the possible results of a simple genetic cross. The diagram below models the Punnett square of a cross between two Tt genotypes.

	T	T
T	TT	Tt
t	Tt	Tt

The Punnett square shows that TT, Tt and tt are the three possible results of the cross. The genotype TT contains two dominant alleles. This genotype is described as homozygous dominant. Homozygous means that the two alleles are the same, and because they are both dominant, the dominant phenotype appears. The table displays two Tt genotypes. This genotype is described as heterozygous because the two alleles are not the same. In the case of a heterozygous result, the dominant phenotype manifests. The final possible result is a tt genotype.

This genotype is described has homozygous recessive. This is the only case in which the recessive trait will manifest. Any simple cross between two heterozygous genotypes will result in a 1 homozygous dominant: 2 heterozygous: 1 homozygous recessive ratio among the inheritance possibilities.

One example of a simple genetic cross in which this type of Punnett square is used is gender. In humans, gender is determined using X and Y alleles. The X allele contains a wide variety of genetic information, including female genetic information. It is recessive. The Y allele contains information relating to male genetic information and is dominant. The Punnett square below shows the possibility that emerges for each human child.

An XX genotype will result in a female child, and a XY genotype will result in a male child. As shown above, the possibility of a child being male or female is fifty percent.

This method works well for very simple genetic crosses; however, most genetic traits are influenced by a number of genes. It is possible to construct a Punnett square which displays the possible crosses of multiple genes. A dihybrid cross; for example, is the cross between two homozygous dominant genes, such as AABB, and two homozygous recessive genes, such as aabb. However, the more genes that are added, the more complex the system becomes. First, it must be determined all of the possible matches between the two types of genes, and then all of those possibilities must be crossed. The Punnett square below demonstrates the cross between an AABB genotype and an AaBb genotype.

	AABB				
		AB	Ab	AB	Ab
AaBb	AB	AABB	AABb	AABB	AABb
	AB	AABB	AABb	AABB	AABb
	aB	AaBB	AaBb	AaBB	AaBb
	aB	AaBB	AaBb	AaBB	AaBb

The Hardy-Weinberg Law says that under these conditions, both phenotypic and allelic frequencies remain constant from generation to generation. This is valid in sexually reproducing populations and is referred to as the Hardy-Weinberg equilibrium.

No mutation
Random mating
No immigration or emigration
Random reproductive success
Large population size

This law proves that natural selection is required for evolution to occur. Darwin's theory of evolution says that in order for evolution to occur, a population must be variable, and that there must be inheritance between generations, while natural selection will make survival and reproductive success non-random.

The conditions set up by the Hardy-Weinberg Law allow for variability and inheritance, but they eliminate natural selection. The fact that no evolution occurs in a population meeting these conditions ultimately proves that evolution can only occur through natural selection.

This also allows us to estimate the effect of selection pressures by studying the difference between actual and expected allelic frequencies or phenotypes. To make these measurements, we need to write a Hardy-Weinberg equation for the frequencies of alleles in the population.

p = the frequency of the dominant allele (represented here by A)
q = the frequency of the recessive allele (represented here by a)

For a population in genetic equilibrium:
$p + q = 1.0$ (The sum of the frequencies of both alleles is 100%)

$$(p + q)^2 = 1$$

so

$$p^2 + 2pq + q^2 = 1$$

The three terms of this binomial expansion indicate the frequencies of the three genotypes:

p^2 = frequency of AA (homozygous dominant)

$2pq$ = frequency of Aa (heterozygous)

q^2 = frequency of aa (homozygous recessive)

Chromosomes in DNA carry genes.

A **Somatic Cell** is a full set of chromosomes (there is a total of 46 chromosomes, thus 46 genes). **Cloning** is done by doing reproduction just using the somatic cell.

Gametes are the reproductive cells (eggs and sperm). Each has exactly one-half a set of normal chromosomes; this is why you need one of each to conceive. Gametes fuse together to make a zygote. A **zygote** is the first part of a human. Only a few genes are on the "y" chromosome. All genes on the "y" are passed on to boys *every time*, but never to girls.

There are some genes that are linked to the sex or gender of the person. **Hemophilia** is passed from mother to son, harmless to the mother but can be fatal to the son. Other disorders that can be passed are **colorblindness** and **muscular dystrophy**.

Animal Breeding

Pure bred dogs are dogs that are only bred with their same type of dog for a favorable trait or look. This can cause health problems for the dogs as sometimes the owners are inbreeding the animals. For example, this is responsible for Saint Bernards having poor hipbones. There are laws against human inbreeding for that same reason because it can cause major health problems and deformations.

Polygenetic: two pairs of genes are used to determine what is passed on.

Rh Factor

Rh factor has to do with your blood. If you are A positive, you are Rh positive. If a husband and wife have different blood types, then it can cause a problem for the baby. If the baby has Rh-positive blood and the mother is Rh negative, it could cause the mother's cells to attack the baby. To fix this, injections are given in the doctor's office or hospital.

Albinism

From the National Organization of Albinism and Hyperpigmentation, here is some information about the disease: "People with albinism have little or no pigment in their eyes, skin, or hair. They have inherited genes that do not make the usual amounts of a pigment called melanin. When both parents carry the gene, and neither parent has albinism, there is a one in four chance at each pregnancy that the baby will be born with albinism. This type of inheritance is called autosomal recessive inheritance." People who have it are sometimes called Albinos. Albinism can affect people from all races.

Tay-Sachs Disease

The National Institute of Neurological Disorders and Strokes published the following information on Tay-Sachs disease:

What is Tay-Sachs disease?
Tay-Sachs disease is a fatal genetic disorder in which harmful quantities of a fatty substance called ganglioside GM2 accumulate in the nerve cells in the brain. Infants with Tay-Sachs disease appear to develop normally for the first few months of life. Then, as nerve cells become distended with fatty material, a relentless deterioration of mental and physical abilities occurs. The child becomes blind, deaf and unable to swallow. Muscles begin to atrophy and paralysis sets in. A much rarer form of the disorder which occurs in patients in their twenties and early thirties is characterized by unsteadiness of gait and progressive neurological deterioration.

Patients with Tay-Sachs have a "cherry-red" spot in the back of their eyes. The condition is caused by insufficient activity of an enzyme called hexosaminidase A that catalyzes the biodegradation of acidic fatty materials known as gangliosides. Gangliosides are made and biodegraded rapidly in early life as the brain develops. Patients and carriers of Tay-Sachs disease can be identified by a simple blood test that measures hexosaminidase A activity. Both parents must be carriers in order to have an affected child. When both parents are found to carry a genetic mutation in hexosaminidase A, there is a 25 percent chance with each pregnancy that the child will be affected with Tay-Sachs disease. Prenatal monitoring of pregnancies is available if desired.

Is there any treatment?
Presently, there is no treatment for Tay-Sachs.

What is the prognosis?
Even with the best of care, children with Tay-Sachs disease usually die by age 5.

Cystic Fibrosis

From the U. S. Department of Health and Human Services, information about Cystic Fibrosis:

"Cystic fibrosis (CF) is a chronic, progressive, and frequently fatal genetic (inherited) disease of the body's mucus glands. CF primarily affects the respiratory and digestive systems in children and young adults. The sweat glands and the reproductive system are also usually involved. On the average, individuals with CF have a lifespan of approximately 30 years. CF-like disease has been known for over two centuries. The name, cystic fibrosis of the pancreas, was first applied to the disease in 1938."

According to the data collected by the Cystic Fibrosis Foundation, there are about 30,000 Americans, 3,000 Canadians and 20,000 Europeans with CF. The disease occurs mostly in whites whose ancestors came from northern Europe, although it affects all races and ethnic groups. Accordingly, it is less common in African Americans, Native Americans and Asian Americans. Approximately 2,500 babies are born with CF each year in the United States. Also, about 1 in every 20 Americans is an unaffected carrier of an abnormal "CF gene." These 12 million people are usually unaware that they are carriers.

CF does not follow the same pattern in all patients but affects different people in different ways and to varying degrees. However, the basic problem is the same-an abnormality in the glands, which produce or secrete sweat and mucus. Sweat cools the body, mucus lubricates the respiratory, digestive, and reproductive systems and prevents tissues from drying out, protecting them from infection.

People with CF lose excessive amounts of salt when they sweat. This can upset the balance of minerals in the blood, which may cause abnormal heart rhythms. Going into shock is also a risk.

Mucus in CF patients is very thick and accumulates in the intestines and lungs. The result is malnutrition, poor growth, frequent respiratory infections, breathing difficulties, and eventually permanent lung damage. Lung disease is the usual cause of death in most patients.

CF can cause various other medical problems. These include sinusitis (inflammation of the nasal sinuses, which are cavities in the skull behind, above and on both sides of the nose), nasal polyps (fleshy growths inside the nose), clubbing (rounding and enlargement of fingers and toes), pneumothorax (rupture of lung tissue and trapping of air between the lung and the chest wall), hemoptysis (coughing of blood), cor pulmonale (enlargement of the right side of the heart), abdominal pain and discomfort, gassiness (too much gas in the intestine), and rectal prolapse (protrusion of the rectum through the anus). Liver disease, diabetes, inflammation of the pancreas and gallstones also occur in some people with CF.

Genes are the basic units of heredity. They are located on structures within the cell nucleus called chromosomes. The function of most genes is to instruct the cells to make particular proteins, most of which have important life-sustaining roles.

Every human being has 46 chromosomes, 23 inherited from each parent. Because each of the 23 pairs of chromosomes contains a complete set of genes, every individual has two sets (one from each parent) of genes for each function. In some individuals, the basic building blocks of a gene (called base pairs) are altered (mutated). A mutation can cause the body to make a defective protein or no protein at all. The result is a loss of some essential biological function and that leads to disease. Children may inherit altered genes from one or both parents.

Diseases such as CF that are caused by inherited genes are called genetic diseases. In CF, each parent carries one abnormal CF gene and one normal CF gene but shows no evidence of the disease because the normal CF gene dominates or "recesses" the abnormal CF gene. To have CF, a child must inherit two abnormal genes; one from each parent. The recessive CF gene can occur in both boys and girls because it is located on non-sex-linked chromosomes called autosomal chromosomes. CF is therefore called an autosomal recessive genetic disease.

The inheritance patterns for the CF gene are shown in the accompanying diagram. Each child, whether male or female, has a 25 percent risk of inheriting a defective gene from each parent and of having CF. A child born to two CF patients (an unlikely event) would be at a 100 percent risk of developing CF.

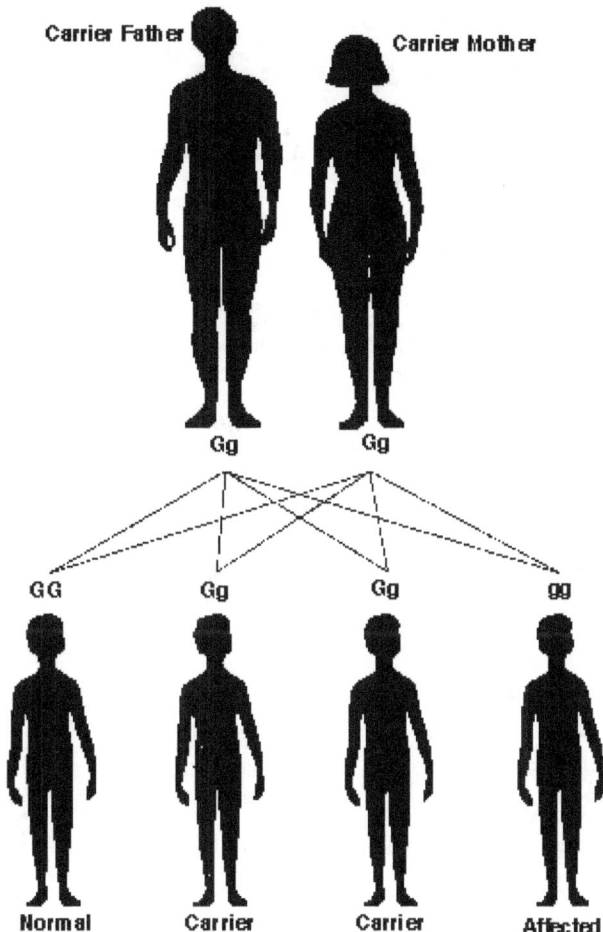

The presence of two mutant genes (g) is needed for CF to appear. Each parent carries one defective gene (g) and one normal gene (G). The single normal gene is sufficient for normal function of the mucus glands, and the parents are therefore CF-free. Each child has a 25 percent risk of inheriting two defective genes and getting CF, a 25 percent chance of inheriting two normal genes and a 50 percent chance of being an unaffected carrier like the parents.

Sickle Cell Anemia

Sickle cell anemia is an abnormal red blood cell that is not round and makes clogs causing oxygen deprivation.

Prenatal Genetics

Amniocentesis is a test of the amniotic fluid to show any genetic problems in the baby.

Genetic engineering is adding desirable traits by inserting genes.

Artificial insemination is when the sperm is taken from a male and injected into the woman for conception.

Artificial implant is when a zygote (fertilized egg) is inserted into a uterus, someone unrelated. This is what is done with surrogate mothers. The mother hosts the baby that has the genes of a different mother and father.

In vitro fertilization is something that is done in the lab where the sperm and egg are combined and then implanted.

The **ovary** is the place where eggs are developed. **Ovulation** is when the egg matures and breaks through the ovary. The sperm and egg meet in the fallopian tube. When the zygote (egg) does not attach to the sides, the lining is expelled. This is a women's menstruation or period. Birth control pills signal to the body that an egg has not attached (even if it has) and the lining along with the egg are expelled.

Yolk sac is filled with food for the embryo.

Amniotic cavity is where the baby develops. The amniotic cavity can be found in any Chordata/land vertebrae animals.

Placenta is the excretion track for the embryo. The mother is eating for two and excreting for two. Placenta is only found in mammals.

Allantois is holding excretion, choosing not to go.

Sometimes people are born with an abnormal number of sex chromosomes. Because in humans, gender is determined by the presence or absence of the Y chromosome, when a person is born with extra X chromosomes, the extra chromosomes develop into an inactive mass called **Barr bodies.** They are named after Murray Barr who discovered them.

Human Biology

Human development begins at fertilization.

Aspects of development are physical development, intellectual development, personality and social development.

Periods of life span:
- Prenatal stages (conception to birth)
- Infancy to toddlers (birth to 3 years)
- Early childhood (3-6 years)
- Middle childhood (6-12 years)
- Adolescence (12-20 years)
- Young Adulthood (20-40 years)
- Middle Adulthood (40-65 years)
- Late Adulthood (65+ years)

There are two different types of twins. **Monozygotic twins** are identical twins. **Dizygotic** twins are fraternal twins.

A gene is something that determines physical traits and is inherited from one or both parents.

A **zygote** is a fertilized ovum or egg.

Neonate is another name for newborn.

Prenatal development is a critical period, which is the first three months in the womb.

During the embryonic period, growth occurs in two directions. The first is **cephalocaudal**, or from the head downward. This means that the head develops faster than the rest of the body. The second type of growth is **proximodistal**, which is from the center (or spine) outward. This means that the vital organs begin to form before the extremities do.

Anoxia is a lack or deficiency of oxygen. This mostly happens during birth itself, by failing to breathe. Cerebral anoxia is more specifically a lack of oxygen to the brain. This can cause permanent damage and is a fairly common problem with infants. Anoxia is a problem in long labors, or if there are complications during birth. Additionally, infants born prematurely or with a very low birth weight have an increased risk of anoxia.

Body cells have 23 pairs. Men have 23 and women have 23 for a grand total of 46 chromosomes.

Down syndrome is caused by one extra chromosome, for a grand total of 47 chromosomes. To avoid this, parents need to conceive their children at younger ages. A woman's chance of having a Down syndrome baby at age 25 is 1 in 2500, and at age 40 the odds are 1 in 100. Men over 55 also have a higher rate of fathering Down syndrome children. Down syndrome is where the child has one extra chromosome, which is number 21.

Mentally retarded is defined as a low IQ and a mental age of about 4 years old. Of all health problems, those who are mentally retarded are the **least likely to have cerebral palsy as well.**

Autism is a lack of responsiveness to other people. Autism is rare in children of either sex in the first 2 ½ years. At birth, cells in the cerebral cortex are not well connected. They are lacking in myelin, which is insulation for nerves.

Rubella is a disease, also called German measles.

Critical period is a specific period in development when a certain event will have the greatest impact. For example, a certain species of bird has a certain period of time when their young can learn to fly. If they do not learn during that time, then they will never learn to fly.

Fetal alcohol syndrome is when babies have been in the womb when the mother was consuming alcohol. The problems include slowed growth, body and face malfunctions, nervous system disorders and mental retardation.

Fetal tobacco syndrome is what a baby can get if their mother smokes while pregnant. Five cigarettes are too many; the child will have problems, with a 50% greater risk of getting childhood cancer. The baby can also be born with a low birth weight.

Teratology is the study of substances which are harmful to prenatal development. These harmful substances are called **teratogens**. Teratogens are most harmful to the fetus and include alcohol, tobacco and other drugs including marijuana and cocaine. Diseases such as HIV, AIDS and rubella are also teratogens, along with pollutants like mercury and radiation. Although the effects of teratogens are numerous and varied, it is nearly impossible to predict what effects will manifest when a teratogen has been present.

Amniocentesis is when a sample is taken from the fluid in the amniotic sac to be tested for various diseases or genetic traits.

For those giving birth, **medicated delivery** is the most popular. Dr. Grantly Dick Read started teaching natural prepared childbirth. **Gentle birth** is when the baby is born in a pool or bath with dim lights. **Cesarean birth** is a surgical delivery.

Ferdinand Lamaze taught women breathing procedures to get through childbirth. This includes panting.

Growth of the brain is not complete at birth. The cerebral cortex is the least developed part of a newborn.

Development after Birth

Neonates have senses. This means that you can startle them in the womb with a loud noise. Their hearing is mostly developed. The eyes are not fully developed when babies are first born. They can distinguish color very early in their development. Newborns can track movement and bright objects.

Although many people believe that infants lack the ability to focus, this isn't true. The muscles involved in eye focus, called ciliary muscles, develop within about two months. After that, the real problem is with visual acuity the ability to see in detail. Visual acuity is determined by the brain's ability to process optical information and the development of the fovea, not the ability of the optical nerves. As the brain and fovea develop, an infant's eyesight develops.

Doctors agree that breast-feeding is the best way to feed your baby.

The "Virginia Apgar" sometimes referred to as the "Apgar Rating" is what doctors use to judge a baby when it is first born. It is rated on the following areas, getting a score of 0-2 in each area:

- Appearance
- Pulse
- Grimace (reflex, irritability)
- Activity
- Respiration

Infants have the capacity to learn. **Habituation**, a simple type of learning, is to get accustomed to something. Some people live near trains. When a train goes by your house every day, and in the night while you are asleep, it is easy to stop hearing the train. You become used to the noise and it does not bother you.

PKU is an enzyme deficiency. 1 in 14,000 will get it and will become mentally retarded if they do not receive treatment. They are tested in the first 3-6 weeks of life. PKU is short for **phenylketonuria.**

SIDS or Sudden Infant Death Syndrome is when there is an apnea, a temporary stoppage of breathing, which causes the baby to die.

The **visual cliff** shows that infants have depth perception. This is a study that was done by having infants placed on a solid, opaque surface. The infant's mothers were placed at the end of the table, where the opaque surface disappeared and glass began. The infants did not want to cross the glass because they could see the distance between where they were and the floor.

Temperament in infants is measured by monitoring:
- Irritability
- Social responsiveness
- Activity level

Object permanence is developed around age two. This means that a child will understand that once you leave the room, you are not gone forever. If you go around a corner, out of their vision, they know you are just around the corner.

According to the psychoanalytic theory, infants develop a secure attachment when a parent pays attention to, and responds appropriately to the needs of the child. This is critical for the emotional development and security of a child.

John Bowlby's attachment theory states that infants need to form at least one strong attachment, such as to a parent, in order to develop normally. He also believed that the attachment held evolutionary purpose. The infant tends to want proximity to the person who cares for them as a form of self-preservation. According to Bowlby's theory, if a child feels that their caregiver is nearby and attentive, they feel confident to explore their surroundings. If not, they feel anxious and over time they become depressed.

Harry Harlow did an experiment with baby monkeys about affection and love. He took baby monkeys away from their mothers while they were still nursing and gave them two pretend monkeys in their cage. One was made of wire and had a bottle. The other was made of cloth but did not have a bottle. The monkeys preferred the one with cloth so much that they clung to it, only interacting with the wire monkey for food.

While it is not uncommon for young children to use different hands for different tasks, consistent hand preference, or hand dominance, begins to manifest between the ages of two and three and is cemented by age six. While being ambidextrous is often considered impressive or unique, it is generally recommended that children have a dominant hand to aid in learning fine motor skills and specialized activities. With cutting, for example, the dominant hand learns to use scissors, while the other hand learns to hold the paper effectively.

Body Essentials

90% of the body is **water**.

Carbohydrates are sugars and starches.

Proteins are sources of essential amino acids.

Fats in general are storage for food. Your diet should have 20-30% taken from fat. Most people in the United States are at 40% or higher. The body combines glycerols with organic acids to make fat.

Vitamins are needed in small amounts. If you don't get enough, you can get **scurvy**. Scurvy is a lack of vitamin C. Another example of the body needing vitamins and minerals is when the body does not get enough calcium or potassium there is an increase in muscle cramps and spasms. Minerals are used to build up specific tissues.

Bacteria

Fermentation is yeast used to make alcohol. The yeast is actually alive and is considered an animal. Lactate fermenters are used with yogurt and cheese without oxygen.

Louis Pasteur was the father of microbiology. Since bacteria was discovered we have developed antibiotics and improved public sanitation.

Fermentation also takes place in the body. Fermentation is the process which breaks down glucose anaerobically, and produces 2 ATP and toxins. The 2 ATP are only a fraction of the energy available in glucose; however, the process is very useful in creating rapid bursts of ATP, such as a person would need when running. The downside is that it also produces toxins. These are carried away by the blood and bonded in the liver. However, in the meanwhile, the buildup causes a burning sensation in the muscles. The process also causes an oxygen deficit, which is why after doing quick exercise like running, a person continues to breathe heavily afterward.

Disease

Malaria is a serious, sometimes fatal disease caused by a parasite. Malaria is spread by mosquitoes that inject the parasite into the person's blood. In the past, DDT was used DDT to exterminate mosquitoes until that was considered harmful.

A **virus** is a protein that mimics RNA and DNA. Viruses act as parasites, fooling the body into thinking they belong there. Some examples include the common cold, flu and measles, AIDS and herpes.

Trichinosis occurs when parasites lay their eggs in the muscles of animals, mostly pork or chicken. A person can be infected by eating the meat.

Cancer happens when genetic material in a cell is damaged. The cell reproduces out of control and makes tumors that can then spread to other parts of the body. There are malignant tumors and benign tumors. Many people believe that benign tumors are not as bad as malignant tumors, but they are still dangerous. Benign tumors are cells restricted to an area; they are still malicious as they show the body is not functioning normally.

Hypertension is when you have high blood pressure. This means that it is more difficult for the heart to pump blood.

Atherosclerosis is when plaque builds up in the arteries and causes blockage.

Leukemia is a form of cancer, when you have too many white blood cells.

Hepatitis C is a disease in which 80% of people who have it experience no symptoms. Those that do experience symptoms may have jaundice, fatigue, dark urine, abdominal pain and nausea. Those affected with Hepatitis C often need liver transplants.

Functions of the Body

The thyroid regulates metabolism and needs iodine to function. Iodine was added to table salt for this purpose.

Adrenal glands give adrenaline to the body to support the "fight or flight" response.

Leukocytes are white blood cells and fight infection. **Erythrocytes** are red blood cells that bring oxygen to the different parts of the body.

There are different types of muscle tissue. They are:

- Epithelial
- Connective- examples of this tissue are bone, cartilage and bone marrow
- Nervous
- Muscle- There are three types of muscle tissue:
    - Smooth- veins and arteries, areas around organs
    - Skeletal- striated, around skeleton, they help you move
    - Cardiac- in the heart, they pump blood

When the diaphragm is filled with air, the position is lowered.

Hepatic Portal Circulation regulates sugar level, amino acids that include the intestines, spleen, pancreas and gall bladder.

When food is consumed by a person, here's what happens to it:

1. Stomach lining absorbs the food
2. Intestines pull out what your body wants from the food
3. Pancreas (which produces insulin and glucagon) gets the food ready to enter the blood stream
4. Liver (cleans the blood) produces bile that helps break down fatty foods
5. Gall Bladder saves up the bile from the liver
6. By the time it gets to the small intestine, all the chemical work has been done, so now the nutrient molecules can be absorbed.
7. Large intestine gets rid of solid waste

Exhaling is an excretion process. It gets rid of hydrogen, carbon and oxygen.
We get hydrogen, carbon and oxygen from sugars. We get nitrogen from proteins. Nitrogen is also excreted. Unlike humans, fish and other aquatic creatures excrete ammonia. They don't use kidneys but excrete through their scales.

Urea is waste from birds and reptiles. It looks like a white paste. Urea is the way they get rid of all their waste to conserve water. They do not create urine like mammals or amphibians. Human waste comes from the breakdown of amino acids in our food that we consume. Mammals and amphibians use their kidneys to create urine.

Birds, lizards, snakes and insects don't get a lot of water. They create uric acid. There is no water in it. A paste is their feces/urine. All creatures that do this lay eggs.

 The Immune System and Immunizations

Memory cells are the body's way of protecting itself against antigens or anything which the immune system thinks is dangerous. Thousands of different types of B cells, or B-Lymphocytes, are produced every day in the bone marrow and are designed to detect certain antigens. Once a B cell finds its particular antigen, the T cells kick in.

There are several different types of T cells. Helper T cells are activated when the B cell finds its antigen. Their job is to tell the B cells to produce antibodies and to activate the cytotoxic T cells. Cytotoxic T cells or "killer T cells" attack the antigen. One of the most important functions of the immune system is to determine what should be attacked and what is part of the body. Suppressor T cells stop the immune system from attacking something that is part of the body. In other words, it suppresses the activity of B cells and cytotoxic T Cells. Once the body has learned how to defeat a specific antigen, it creates memory B and T cells. These cells have long lives, and their job is to remember how to defeat that antigen in case it is ever present again. This is why some diseases can only be caught once, such as chicken pox. Once a person has had it, the memory cells know how to defeat it if it sees it again.

Human Immunodeficiency Virus or HIV is a disease which helps show why helper T cells are so important to the immune system. HIV attacks the helper T cells, which are what activate cytotoxic T cells and signal the B cells to make antibodies. When there is not a sufficient number of helper T cells, the immune system has a much more difficult time fighting infections.

Vaccines cause the body to create B cells which fight against a specific virus. While these B cells can last for years at a time, it is necessary to receive booster injections to create more when they wear out. These booster shots must be timed carefully, because if they are administered too early, the existing B cells will destroy the injection before more cells are created, and if they are administered too late it leaves a chance of a person catching the virus. Another reason to receive booster injections is because scientific improvement makes it so that the vaccines become stronger. Also, with illnesses such as the flu, the virus is always changing and mutating, so it is best to keep up with the most recent shots which protect against the most prevalent strains.

 Temperature Regulation in Animals

Ectotherms are animals whose body temperature is regulated by environment. For example, cold-blooded animals and insects.

Endotherms are animals who regulate their body temperature. Birds and mammals do this. Normal body functions require a certain temperature.

The **endocrine system** is a collection of glands that produce and store hormones. The hormones regulate metabolism, growth and sexual development. These glands release the hormones directly into the bloodstream to be sent to the rest of the organs in the body.

When there is an **organ transplant**, the white blood cells may attack the new organ because they see the organ as an invader.

Cornea transplant is the most successful transplant. The cornea is a protective covering for the eye and has no blood flow.

Echolocation, also known as ultrasound is what bats, whales and dolphins use to navigate.

Insects

Pheromones or scents are used by insects to communicate with each other. They leave a trail behind them to follow each other and give warnings or give other messages. This can also be used to stake out territory and to signal sexual arousal.

Bees use a dance language where they dance in a figure eight pattern. They also wiggle their bottom to make other signals. They are social insects that live in colonies or hives and have different jobs such as a worker bee or a queen bee.

Spiders and grasshoppers live by themselves.

Plants

Plant cells have a cell wall. Animal cells do not.

When a plant bends towards the light it is called **positive phototropism.** This bending is caused by **auxin.** When a plant is exposed to sunlight on one side, auxin gathers in the shaded side of the stem and causes the cells to elongate. This one sided elongation causes the plant to tilt toward the sun.

The **vascular cambium** goes all the way up and to the veins of the leaves. It sends out minerals and water to all parts of the tree. **Transpiration** is the evaporation of water from plants from when the stomata in the cell are open during photosynthesis. **Xylem** is the tube part of the plant that transports the nutrients and minerals.

Stoma is the opening in leaves where transpiration takes place. Water evaporates from the stoma sucking up the xylem into the leaves. During the autumn or fall, the vascular cambium dries up and dies. In the spring, a new vascular cambium grows; that is why trees get fatter. Trees grow from the top up.

Waxy coating on plants and the stoma opening help in hot climates to keep water in. Animals and bees pollinate the flowers and plants by carrying the pollen on their bodies as they go from plant to plant.

Krebs cycle

The Krebs cycle, also called the citric acid cycle, is the process through which the cell converts carbohydrates, fats and proteins into energy. The cycle begins with acetyl-CoA combining to form citric acid. It moves through a number of other conversions, each of which release energy, and ends with a bond forming acetyl-CoA once more. The Krebs cycle occurs in all oxygen dependent cells.

Calvin cycle

The Calvin cycle takes place in the stroma of photosynthetic plants. The process is used to create glucose by using carbon dioxide. It is a phase of photosynthesis, even though it doesn't require light. There are three main stages. The first is called fixation. This is when the carbon dioxide is bonded with a protein which occurs naturally in the body. The second stage is reduction. In this stage, the carbon dioxide and protein combination is transformed into a carbohydrate using NADHP and ATP. The third stage is called regeneration. In this stage, ATP is used to convert the carbohydrate into glucose.

The lifecycle of a fern

The lifecycle of a fern is interesting because it includes what is called an alternation of generations. This means that it reproduces in two stages, both sexual and asexual. In its first stage, which is asexual, the fern is called a sporophyte. As a sporophyte, the fern produces leaves. These leaves produce spores, which grow into gametophytes. This is where the sexual part of development begins. The gametophyte grows both male and female organs which produce male and female sex cells called gametes. These join to form a single cell, which grows into a new fern over time.

Flower Reproductive Parts

A- Stigma, sticky and traps pollen grains
B- Style, tube leading to the ovary
C- Ovary
D- Petal
E- Sepal, the outer flower structure
F- Receptacle
G- Stalk, supports the supporting the flower
H- Filament, supports the anther
I- Anther, produces pollen grains that contain sperm cells

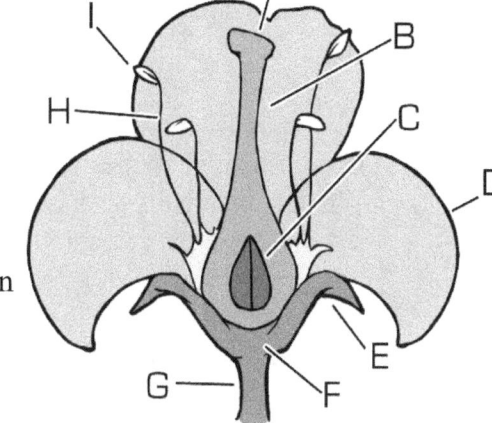

Stamen- the male organ
- Anther
- Filament

Pistil- the female organ
- Stigma
- Style
- Ovary
- Ovules

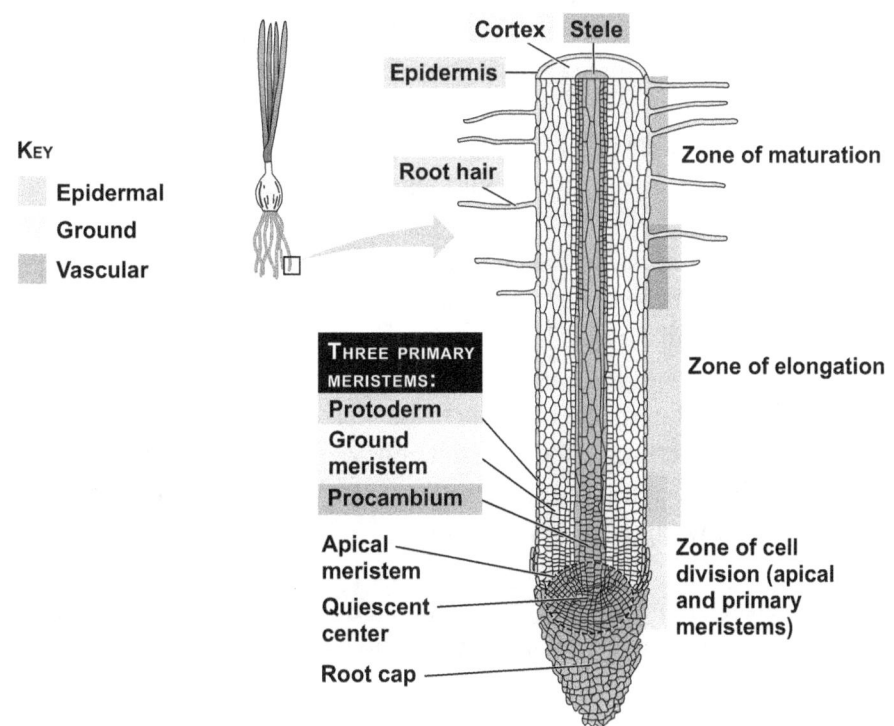

Research

To study the way that people grow, learn, adapt and interact with others, psychologists use a standardized method so that other people in the scientific community can understand their findings and agree on research.

Scientists use a specific vocabulary to conduct their research. A **participant** is a person that a scientist studies in their experiment. They can also be referred to as a subject. When a scientist is performing an experiment on an animal, they are also referred to as a subject or a participant.

When scientists want to study an entire city, culture, or population, they will use a sample. A **sample** is a small collection of subjects. The number of people you need to participate to make the sample the most accurate is statistically generated based on the amount of the population.

Everything that a scientist measures and studies is called a **variable**. For example, if you were conducting research on insomnia, you would have variables which include the amount of time it takes a person to fall asleep, how much caffeine they ingest, how much alcohol or drugs they ingest, what distractions are in the room, etc.

Research must meet four main tests:
1. Research must be **replicable**. Another scientist, given the information regarding the experiment, should be able to reproduce the experiment with the same results. This is how the scientific community accepts or rejects new theories. If the experiment can be reproduced several times by different people in different organizations or locations, it lends to its credibility. This means that the theory must be quantitative or measurable and not qualitative. Qualitative means that something is similar in structure or organization, but it cannot be measured in numerical terms.

2. The research must be **falsifiable**. This means that a theory has to be stated in a way that can be rejected or accepted. Think of it as asking a yes or no question. Is smoking bad for you? The answer is yes or no, and can be proven. This could be stated as "smoking is bad for you because it contains carcinogens." This is a falsifiable statement. It needs to be stated this way so that it can be proved or disproved. If a researcher does not consider all the evidence, but ignores the information that does not prove their theory and accepts the information that proves it, they are showing **confirmation bias**.

3. The research must be precisely stated and conducted. A theory needs to be stated precisely so it can be replicated. Scientists use operational definitions to state ex-

actly how a variable will be measured. For example, a researcher studying birth order may notice that children who are the oldest of several siblings tend to be more responsible as adults and parents. The researcher may conclude that this is because they have experienced more time nurturing and caring for younger siblings. In our birth order study, the scientist needs to find a way to measure "responsible". He or she might decide that they will use the individual's credit reports as an operational variable to show responsibility.

4. Researchers must use the most logical, simplest explanation possible as an answer to their theory. This is also called the **principle of parsimony** or **Occam's razor**.

THE SCIENTIFIC METHOD

The scientific method is the accepted way to conduct research. It contains several steps. First, a researcher created a hypothesis. This is the testable idea. Second, information is gathered through an experiment or research. This information either proves or disproves the theory which leads to the third step, refining the theory. At this point, it may be necessary to start the experiment all over, having applied the new information learned in the experiment. The fourth and final stage is developing a theory. Again, at this point, it is necessary to test the theory through the scientific method. Once a theory has been proved successfully by reputable researchers, and the more times it is reproduced, the more credibility it has.

RESEARCH METHODS

Information for a theory or experiment can be gathered several ways.

Case Study

In a case study, a single individual (subject) is intensely studied. The researcher gets data through personal interviews with the subject, its employees, neighbors, contacts, etc., and by reviewing documentation or records (i.e., medical history, family life, etc.). Other sources for information are testing and direct observation of the subject.

Survey

A survey is a great way to get information about a specific type of information. For example, a survey would work well to measure performance in an office environment. These can be aggregated and used to improve employee performance. Usually with a survey, questionnaires are given out to participants who are then asked to answer questions to the best of their ability. When a participant fills out a survey themselves *about* themselves, it is called self-report data. This information can possibly not be as reliable as other research methods because subjects may be dishonest with their answers. For example, the question "Are you ever late to work?" may have respondents answering "no" when in fact, they are late but either do not remember that or are dishonest to avoid punishment or negative information about themselves. Many give the answers they feel that researchers (or themselves) want to hear instead of the truth.

Naturalistic Observation
Jean Piaget extensively used natural observation to study children. Naturalistic observation is when a researcher observes and studies subjects without interacting or interfering with them. Piaget observed the behavior of children playing in the schoolyard to assess developmental stages. Another example well known to television viewers of the series "Star Trek" involves "The Prime Directive." This is the most perfect version demonstrated (in fiction) of naturalistic observation. In the show, the researchers had the ability to view and study human cultures without being known to the subjects because of their advancements in technology. In the series, it was a great violation to interact with and impact the development of these cultures and societies.

Laboratory Observation
Laboratory observation is conducted in a laboratory environment. This method is selected to monitor specific biological changes in individuals. In a lab setting, expensive and sophisticated machinery can be used to study the participants. Sometimes one-way mirrors are used to observe the participants.

Psychological Tests
Psychological tests give information about participants. Some of the more common include standardized tests such as the Minnesota Multiphasic Personality Inventory also known as the MMPI (a personality test), aptitudes, interests, etc. A participant's score is then compared to the norms for that test. A test is valid if it measures what it is supposed to. For example, a test on depression will be able to measure a person's depression. If it cannot, then the test is not valid. Content validity is applied when a test measures something with more than that one facet. For example, a test for overall cooking skills would not be valid if it only tested baking cakes and not other skills such as grilling meat or making soup.

Cross Sectional Studies
When people of different ages are studied at one particular time, it is called a cross sectional study because you have a cross section of the population or demographic that you want to study.

Longitudinal Studies
Longitudinal studies are when people are followed and studied over a long period of time and checked up on at certain points. These are best used to study the development of certain traits and track health issues. An example of a longitudinal study would be: 600 infants that were put up for adoption were tracked for several years. Some infants were adopted, some returned to the birth mothers and some were put into foster care. Which group adjusted the best and why?

Correlation Research
Correlation research is used to show links between events, people, actions, behaviors, etc. Correlation research does not determine the causes of behavior but is linked to statistics. Causation is the cause of something. Correlation is not causation. This is an example of FAULTY, incorrect causation: a child eats an ice cream three times a week. This child scores well on school aptitude tests. It is determined that eating ice cream will make you smarter and do better on tests. There are additional factors or many others including socioeconomic status resulting from educated parents who genetically pass on their aptitude for school as well as their influence on the importance of school. In this situation, it is most likely the parents who contribute to the child's aptitude scores.

When conducting a survey and you have completed compiling the data, you will be able to measure the correlation between certain traits and variables tested. A correlation coefficient measures the strength between the two variables. A correlation coefficient is a number between -1 and +1.

A positive correlation means that when one variable increases, the other variable increases as well. For example, the more a couple fights, the more likely they are to get a divorce.

When one variable increases and the other variable decreases, it is called a negative correlation. An example of this would be babies that are held by their caregivers tend to cry less. When the amount of time they are held goes up, the time they cry goes down.

The higher the number of the correlation coefficient, the stronger the correlation. A +0.9 or -0.9 shows a very strong correlation because the number is closest to a whole positive number 1 or a whole negative number 1. A weak correlation is a +0.1 or a -0.1. A correlation of zero shows that there is no relationship between variables.

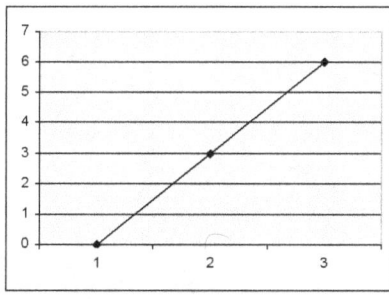
Positive correlation

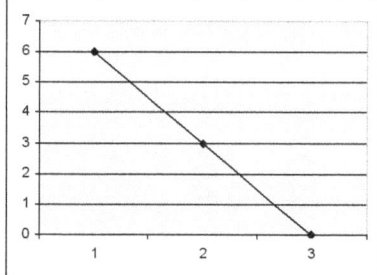
Negative correlation

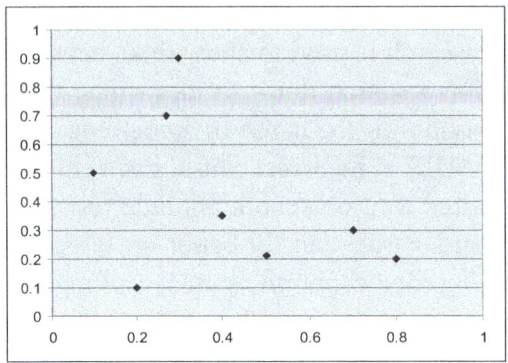

No correlation (above)

Census

A census is a collection of data from all cases or people in the chosen set. Usually, the most common form of a census would take place within an entire school or state. This means that every person of that school or state must be included. Censuses are usually not performed because they are so expensive. A census is valuable because it gives an accurate representation. To save time and money, survey companies will ask 1000 people or so (remember, the number changes based on the amount of people to be surveyed. A good rule of thumb is 10%). This is called sampling. For example, a recent census shows that the single person is the fastest-growing household type. So basically, a sample is a set of cases of people randomly chosen from a large group. The sample is to represent the group. The larger the sample, the more accurate the results.

READING CHARTS AND GRAPHS

Charts and graphs are easy ways to display information and make it easily readable.

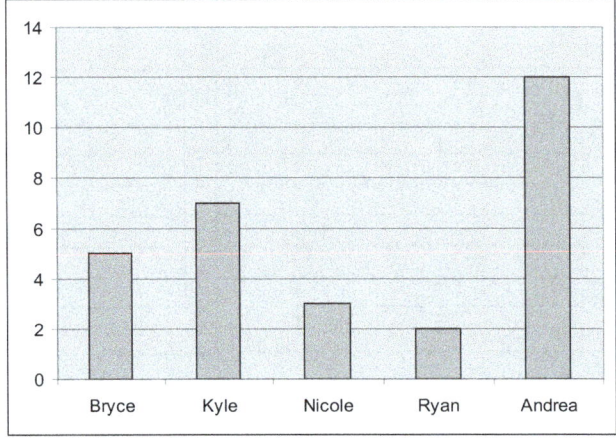

The above is a bar chart which shows five student's hours per week that they practice the piano. Are you able to tell who has the most hours and who has the least? How many hours per week does Nicole practice?

Name	Hours Per Week
Andrea	12
Bryce	5
Kyle	7
Nicole	3
Ryan	2

EXPERIMENTS

In experiments, a researcher manipulates variables to test theories and conclusions. Each experiment has independent and dependent variables. This is how researchers test cause and effect links and relationships.

The independent variable is the variable that researchers have direct control over. The dependent variable is then observed by the researcher.

In experiments, there are usually two groups of participants. One group is the experimental group and one group is the control group. The most common example is in medical trials. Let's say there is a trial run of a new diet drug. The researcher will split the group randomly in two. Group 1 will receive the diet pill that is being tested. Group 2 will receive a placebo pill. The placebo pill is simply a sugar pill. Group 2 will not know that they are not receiving the real drug. This allows the researchers to study the true effectiveness and side-effects of the pill. When the people are assigned to a group randomly, it is called **random assignment**. This particular experiment was a single-blind experiment. A **double-blind** experiment is when none of the doctors, researchers and participants know who is getting the real drug. It is assigned by computer or an independent individual where it is kept confidential until the conclusion of the study.

When a participant starts to feel the effects of the drug but is *actually* taking a sugar pill or placebo it is called the **placebo effect**.

It is very important to avoid bias in research. Bias is the distortion of the results. Common types of bias include the sampling bias, subject bias and researcher bias. The placebo effect is an example of subject bias. Experimenter or researcher bias is avoided by conducting a double-blind experiment.

There are some disadvantages to experiments. They cannot be used to study everything. There are officially defined rules how humans and animals must be treated with the experiment. In an infamous experiment by psychologist Stanley Milgram, subjects were told that they were giving painful electric shocks to other people when in reality they were not. Some people consider this experiment unethical because it caused the participants emotional discomfort.

Researchers must get consent from their participants before conducting the experiment. Informed consent means that the participants must know the content of the experiment and be warned of any risk or harm.

INTERPRETING DATA

The "Mean" is also referred to as the Arithmetic Mean.

The Arithmetic Mean is calculated by finding the **sum** of all "N" values and dividing by N. This is the general formula to calculate the arithmetic mean:

Example of Arithmetic Mean:

What is the arithmetic mean of the numbers 10, 8, 6, 12, and 9?

$X = (10 + 8 + 6 + 12 + 9) = 45$ and $N = 5$ because there 5 numbers in the list. Therefore, = **9**

What is the mode? The mode is the number that appears the most number of times in the sample data.

For example, the mode for the following set of data {10, 12, 14, 10, 8, 10, 7} is **10** because 10 appears the most number of times in the data set.

What is the median? The median is the middle value or the arithmetic mean of the middle values. The following example will demonstrate how to find the median.

Sample Data = {3, 3, 5, 6, 7, 9, 10, 12, 14}

The median = 7 because it has four numbers to the left of it and four numbers to the right of it in the sample data.

> NOTE: The data MUST be in numerical order to find the median. For example, if you are given a sample data set of {5, 6, 12, 3, 3, 10, 7, 14, 9} you must first put it in numerical order: {3, 3, 5, 6, 7, 9, 10, 12, 14} to find the correct median of 7.

Find the median for the following set of numbers: {2, 3, 4, 5, 6, 7}

Since there is an even number of numbers in the data set you will need to find the arithmetic mean of the two middle numbers 4 & 5. (4+5)/2 = 4.5. Therefore, the median is 4.5.

What is the range? The range is simply the difference between the largest data value and the smallest data value. For example, the range in the set of {2, 4, 7, 8, 10, 12, 24} is 24-2 = **22.**

Sample Test Questions

1) The primary and universal carrier of chemical energy in living beings is the ATP, this was suggested by

   A) Albert Lehninger
   B) Fritz Lipmann
   C) Karl Lohmann
   D) Y. Subbarow
   E) None of the above

The correct answer is C:) Karl Lohmann.

2) After a person moves to a new area, they often hear airplanes flying overhead. After a week or two, they rarely notice them. What is this an example of?

   A) Auditory cliff
   B) Telegraphic speech
   C) Habituation
   D) Presbycusis
   E) Phoneme

The correct answer is C:) Habituation.

3) The reason why vegetatively reproducing crop plants are best suited for maintaining hybrid vigor is that

   A) They are more resistant to diseases
   B) Once a desired hybrid has been produced there are few chances of losing it
   C) They can be easily propagated
   D) They have a longer life span
   E) None of the above

The correct answer is B:) Once a desired hybrid has been produced there are few chances of losing it.

4) Where is the pollen found in a flower?

   A) Stamen
   B) Sepals
   C) Anther
   D) Carpels
   E) Stigma

The correct answer is C:) Anther. Pollen is found in the anther, where it is formed in pollen sacs.

5) What is the name for the process used to create glucose from carbon dioxide?

   A) Calvin cycle
   B) The NADPH process
   C) Glycolysis
   D) Krebs cycle
   E) Photosynthesis

The correct answer is A:) Calvin cycle. This reaction takes place in the stroma of photosynthetic plants. The Calvin cycle is a part of photosynthesis; however, it isn't the entire process.

6) What is the name for the process used to convert carbohydrates, fats and proteins into energy?

   A) Calvin cycle
   B) The NADPH process
   C) Glycolysis
   D) Krebs cycle
   E) Photosynthesis

The correct answer is D:) Krebs cycle. The Krebs cycle is important to metabolism because it is how the body breaks down carbohydrates to produce energy.

7) In a strand of DNA, what would be paired with A-G-C-C-T-A?

   A) C-T-A-A-G-C
   B) T-C-G-G-A-T
   C) G-A-T-T-C-G
   D) A-G-C-C-T-A
   E) Cannot be determined

The correct answer is B:) T-C-G-G-A-T. This can be determined because adenine always pairs with thymine and guanine always pairs with cytosine.

8) "Trehalose" is a disaccharide formed by two glucose units and is a

   A) Reducing sugar
   B) Non-reducing sugar
   C) Reducing aldose
   D) Non-reducing ketose
   E) None of the above

The correct answer is B:) Non-reducing sugar.

9) Cellulose is a glucose polymer with

   A) p-1, 4 linkage
   B) ot-1, 4 linkage
   C) a-1, 6 linkage
   D) p-1, 4 and 1, 6 linkage
   E) None of the above

The correct answer is B:) ot-1, 4 linkage.

10) What is the term referring to the study of the evolutionary history of organisms?

   A) Analogy
   B) Systematics
   C) Homology
   D) Parsimony
   E) Phylogeny

The correct answer is E:) Phylogeny. The study of phylogenics includes the study of different primitive characteristics and derived characteristics to determine line of descent.

11) Four pyrrole rings in heme are joined through Methylidyne Bridge, so to form

   A) Protoporphyrin
   B) Porphin
   C) Porphyrin
   D) Hemoglobin
   E) None of the above

The correct answer is C:) Porphyrin.

12) The pitch of the helix having length of 34 A° is a

   A) A form of DNA
   B) B form of DNA
   C) Z form of DNA
   D) X form of DNA
   E) None of the above

The correct answer is B:) B form of DNA.

13) Where does food go after the stomach?

   A) Transverse colon
   B) Ileum
   C) Duodenum
   D) Large intestine
   E) Esophagus

The correct answer is C:) Duodenum. In the duodenum, the food is completely broken down. For example, from proteins to amino acids, complex sugars to simple sugars, and fats into fatty acids.

14) Radio-tracer technique shows that DNA is in

   A) Multi-helix stage
   B) Double-helix stage
   C) Single-helix stage
   D) Quadra-helix stage
   E) None of the above

The correct answer is B:) Double-helix stage.

15) What is the cotyledon?

   A) The leafy portion of a plant's embryo.
   B) The first leaves present on a growing plant.
   C) The tubes which carry water and nutrients through the plant.
   D) The first root which grows on a plant.
   E) The technical name for a seed.

The correct answer is A:) The leafy portion of a plant's embryo. These wither after the first true leaves appear.

16) What are the mostly diverse molecules in the cell?

   A) Proteins
   B) Mineral salts
   C) Carbohydrates
   D) Lipids
   E) None of the above

The correct answer is A:) Proteins.

17) Choose the correct enzyme-substrate pair

   A) Rennin-caesin
   B) Maltase-lactose
   C) Protein-amylase
   D) Carbohydrate-lipase
   E) None of the above

The correct answer is A:) Rennin-caesin.

18) Which of the following groups are all polysaccharides?

   A) Maltose, lactose and fructose
   B) Glycogen, cellulose and starch
   C) Glycogen, sucrose and maltose
   D) Sucrose, glucose and fructose
   E) None of the above

The correct answer is B:) Glycogen, cellulose and starch.

19) Which purine base is found in RNA?

   A) Uracil
   B) Guanine
   C) Cytosine
   D) Thiamine
   E) None of the above

The correct answer is B:) Guanine.

20) When one molecule of ATP is disintegrated, what amount of energy is liberated?

   A) 7 kcal
   B) 38 kcal
   C) 4.5 kcal
   D) 8 kcal
   E) None of the above

The correct answer is C:) 4.5 kcal.

21) This phenomena occurs after a species is nearly extinct and the gene pool is significantly reduced. This causes certain characteristics to disappear within a species, while others become identical amongst members. What is this called?

   A) Founder's effect
   B) Operant conditioning
   C) Mutualism
   D) Bottleneck effect
   E) None of the above

The correct answer is D:) Bottleneck effect. Due to a similarity in cheetah enzymes, it is speculated that they were once subject to the bottleneck effect.

22) When paternal and maternal chromosomes change their materials with each other in cell division, this event is called

   A) Dyad-forming
   B) Crossing-over
   C) Synapses
   D) Bivalent-forming
   E) None of the above

The correct answer is B:) Crossing-over.

23) Lysosomes are rich in

   A) Nucleic acids
   B) Carbohydrates
   C) Hormones
   D) Hydrolytic enzymes
   E) None of the above

The correct answer is D:) Hydrolytic enzymes.

24) The term 'meiosis' was given by

   A) Flemming
   B) Johansen
   C) Knoll and Ruska
   D) Farmer and Moore
   E) None of the above

The correct answer is D:) Farmer and Moore.

25) When water enters in roots due to diffusion, it is termed as

   A) Osmosis
   B) Endocytosis
   C) Active absorption
   D) Passive absorption
   E) None of the above

The correct answer is D:) Passive absorption.

26) The genetic drift in which rare alleles occur at a higher frequency when a population is isolated is called what?

   A) Founder's effect
   B) Operant conditioning
   C) Mutualism
   D) Bottleneck effect
   E) None of the above

The correct answer is A:) Founder's effect. Because of founder's effect, the Amish of Lancaster County, Pennsylvania have a higher percentage of people with polydactylism than in most populations.

27) Which is a typical example of feedback inhibition?

   A) Allosteric inhibition of hexokinase by glucose-6-phosphate
   B) Sulpha drugs and folic acid synthesizer bacteria
   C) Reaction between succinic dehydrogenase and succinic acid
   D) Cyanide and cytochrome reaction
   E) None of the above

The correct answer is A:) Allosteric inhibition of hexokinase by glucose-6-phosphate.

28) In what way is fermentation a useful process?

   A) It is the most efficient process of breaking down glucose to use the full amount of available energy.
   B) Fermentation is useful because it creates rapid bursts of ATP.
   C) It is the only process of breaking down glucose which does not produce toxins.
   D) It occurs anaerobically and can extend a person's life if they are suffocating.
   E) Fermentation is not a useful process. It is inefficient and produces toxins.

The correct answer is B:) Fermentation is useful because it creates rapid bursts of ATP. While it's true that fermentation produces toxins and causes a temporary oxygen deficit, it is useful when doing anaerobic activities.

29) At the end of glycolysis, six carbon compounds ultimately change into

   A) Pyruvic acid
   B) Acetyl Co-A
   C) ATP
   D) Ethyl alcohol
   E) None of the above

The correct answer is A:) Pyruvic acid.

30) What is suggested by the appearance of gill arches in the stages of development of the embryos of all vertebrates?

   A) Fish, amphibians, reptiles, birds and mammals are probably descendants of a common ancestor.
   B) Invertebrates probably descended from a common ancestor since they do not have the gill arches.
   C) Mammals probably do not share a common ancestor with fish because they have little in common.
   D) Fish, amphibians and reptiles likely descended from a common ancestor, and birds and mammals likely descended from a different common ancestor.
   E) Nothing is suggested by the appearance of gill arches. It is simply a common occurrence.

The correct answer is A:) Fish, amphibians, reptiles, birds and mammals are probably descendants of a common ancestor. Fish, amphibians, reptiles, birds and mammals are all vertebrates, and because they show gill arches in the stages of development, they likely share a common ancestor and have just evolved differently.

31) Which of the following contains natural silk?

   A) Magnesium
   B) Potassium
   C) Nitrogen
   D) Phosphorus
   E) None of the above

The correct answer is C:) Nitrogen.

32) Which of the following products are obtained by anaerobic respiration from yeast?

   A) Beer and wine
   B) Alcohols
   C) CO2
   D) All of these
   E) None of the above

The correct answer is D:) All of these.

33) Which of the following best describes a habitat?

   A) A group of mammals which reside in close proximity.
   B) All of the living organisms which are in an area.
   C) A particular area of a community in which a population resides.
   D) A specific species which a researcher wishes to study.
   E) None of the above

The correct answer is C:) A particular area of a community in which a population resides. For example, underground, in the trees or in tall grass.

34) What is common among amylase, rennin and trypsin?

   A) These are all proteins
   B) These act at a pH lower than average
   C) These are all proteolytic enzymes
   D) These are produced in stomach
   E) None of the above

The correct answer is A:) These are all proteins.

35) In which biome would maple, beech, oak, hickory, basswood, ash and elm trees be found?

   A) Arctic tundra
   B) Taiga
   C) Grassland
   D) Temperate deciduous forest
   E) None of the above

The correct answer is D:) Temperate deciduous forest. The average rainfall in a temperate deciduous forest is 30 to 60 inches annually, with an average temperature of 50 degrees Fahrenheit.

36) Where is the electron transport chain located?

   A) Endoplasmic reticulum
   B) Golgi apparatus
   C) Mitochondria
   D) Ribosomes
   E) Cell membrane

The correct answer is C:) Mitochondria. The mitochondria uses the electron transport chain to produce ATP.

37) The end products of fermentation are

   A) CO2 and acetaldehyde
   B) CO2 and CaH5OH
   C) O2 and C2H5OH
   D) CO2 and O2
   E) None of the above

The correct answer is B:) CO2 and CaH5OH.

38) Which typical stage is known for DNA replication?

   A) G2 phase
   B) G, phase
   C) S-phase
   D) Metaphase
   E) None of the above

The correct answer is C:) S-phase.

39) Which of the following are produced through the electron transport chain?

   A) ATP
   B) NADH
   C) Water
   D) Both A and C
   E) A, B, and C

The correct answer is D:) Both A and C. Energy is released as the electrons move through the chain, and is used to create ATP, so ATP is produced. The final stage is when the electrons are transferred to oxygen which combines with hydrogen to form water.

40) Centromere is a part of

   A) Ribosomes
   B) Endoplasmic reticulum
   C) Chromosome
   D) Mitochondria
   E) None of the above

The correct answer is C:) Chromosome.

41) A parasite that only multiplies in the body of a host is called

   A) Simple parasite
   B) Complex parasite
   C) Facultative parasite
   D) Obligate parasite
   E) None of the above

The correct answer is D:) Obligate parasite.

42) What are the four types of consequences related to operant conditioning?

   A) Something good is introduced
   B) Something good is taken away
   C) Something bad is introduced
   D) Something bad is taken away
   E) All of the above

The correct answer is E:) All of the above. The idea behind operant conditioning is to create an association between an action and a response or consequence. Answers A through D all list possible consequences.

43) A virus that attacks blue green algae is known as

   A) Bacteriophage
   B) Rhodophage
   C) Cyanophage
   D) Zoophages
   E) None of the above

The correct answer is C:) Cyanophage.

44) Which of the following plants are NOT representative of an arctic tundra?

   A) Lichens
   B) Mosses
   C) Sedges
   D) Dwarf shrubs
   E) All of the above are representative of an arctic tundra

The correct answer is E:) All of the above are representative of an arctic tundra. Lichens, mosses, sedges and dwarf shrubs are representative plants of an arctic tundra.

45) An integrated product of viral DNA and host cell DNA is called a

   A) Provirus
   B) Retrovirus
   C) Geno-virus
   D) Antivirus
   E) None of the above

The correct answer is A:) Provirus.

46) Any strain of bacteria that have the chromosomal F-Factor (fertility factor) is called

   A) Super male
   B) Hfr (High Frequency of Recombination)
   C) Sterile male
   D) Impotent female
   E) None of the above

The correct answer is B:) Hfr (High Frequency of Recombination).

47) Which of the following correctly lists the phases of the Calvin cycle in their correct order?

   A) Reduction, regeneration, fixation
   B) Regeneration, reduction, fixation
   C) Reduction, fixation, regeneration
   D) Fixation, regeneration, reduction
   E) Fixation, reduction, regeneration

The correct answer is E:) Fixation, reduction, regeneration. Through the Calvin cycle, carbon dioxide is converted into glucose.

48) The protein coat of a virus is generally named as

   A) Capsid
   B) Capsomere
   C) Lysocoat
   D) Onco-cap
   E) None of the above

The correct answer is A:) Capsid.

49) What type stages of growth are included in the lifecycle of a fern?

   I. Phototrophic
   II. Sexual
   III. Asexual
   IV. Adaptative

   A) III and IV only
   B) II and III only
   C) I and IV only
   D) I, II and IV only
   E) I, II, III, and IV

The correct answer is B:) II and III only. The lifecycle of a fern is unique because it includes both sexual and asexual stages.

50) The genes that are constantly required for various cellular activities are known as

   A) Housekeeping genes
   B) Constitutive genes
   C) Both
   D) Regulatory genes
   E) None of the above

The correct answer is A:) Housekeeping genes.

51) What is imprinting?

   A) Learning which occurs at a specific phase of life.
   B) The ability of the mind to process visual images.
   C) Learning through creating an association between an action and a response or consequence.
   D) The movement of molecules to an area of high concentration to low concentration.
   E) Creating an exact genetic replica of an organism.

The correct answer is A:) Learning which occurs at a specific phase of life. For example, animals believing the first creature they see to be their parent is imprinting because it is specific to their first few moments of life.

52) What is it called when a plant reproduces using both sexual and asexual methods?

   A) Bi-methodical reproduction
   B) Sympatric speciation
   C) Bioremediate reproduction
   D) Alternation of generations
   E) None of the above

The correct answer is D:) Alternation of generations. One example of an organism to which this applies is the fern.

53) A population of genetically identical cells formed by vegetative cell division is called

   A) Exon
   B) Intron
   C) Clone
   D) Genome
   E) None of the above

The correct answer is C:) Clone.

54) What do extra sex chromosomes develop into?

   A) Prions
   B) Nucleotides
   C) Gelatinous Masses
   D) Barr bodies
   E) It is not possible to survive with excess sex chromosomes, so they are not named.

The correct answer is D:) Barr bodies. Barr bodies are inactive chromosomes.

55) The study of the science of aging is known as

   A) Gerontology
   B) Deterioration
   C) Calcification
   D) Epidemiology
   E) None of the above

The correct answer is A:) Gerontology.

56) How is biomass determined?

   A) If an organism has mass, and is living, it is considered to be one unit of biomass.
   B) By multiplying the number of animals by their weight.
   C) By multiplying the number of adult animals by their mass, because young animals don't matter.
   D) By determining the number of animals and multiplying by ten.
   E) None of the above

The correct answer is B:) By multiplying the number of animals by their weight. The greatest amount of biomass and stored energy, are always at the lowest levels of the biomass pyramid.

57) In order, the stages of mitosis are prophase, prometaphase, metaphase, anaphase, telophase, cytokinesis. During which stage do the chromosomes align in the middle of the nucleus?

   A) Prophase
   B) Prometaphase
   C) Metaphase
   D) Interphase
   E) Interphase

The correct answer is C:) Metaphase. This alignment helps the chromosomes to divide properly in the anaphase stage, so each new cell gets a copy of each chromosome.

58) Which of the following is true of both prokaryotic AND eukaryotic cells?

   A) They contain a nucleus
   B) They contain ribosomes
   C) They contain a membrane and cell wall
   D) They are usually multicellular
   E) None of the above

The correct answer is B:) They contain ribosomes. Both types of cells contain genetic material, a cell membrane, ribosomes, vacuoles and vesicles, but most other structures are different.

59) The reflectivity percentage of incident light on the earth called

   A) Irrefraction
   B) Albedo
   C) Tornado
   D) Eradiation
   E) None of the above

The correct answer is B:) Albedo.

60) A small egg

   A) Ovary
   B) Ovule
   C) Stamen
   D) Pistil
   E) None of the above

The correct answer is B:) Ovule. An ovule is a small outgrowth of the ovary of a seed plant after being fertilized develops into a seed.

61) Kranz anatomy is seen in

   A) Mangifera indica
   B) Zea mays
   C) Citrus indica
   D) Euphorbia hirta
   E) None of the above

The correct answer is B:) Zea mays.

62) In which biome are evergreen conifers the most prominent plant?

   A) Taiga
   B) Temperate deciduous forest
   C) Rainforest
   D) Arctic Tundra
   E) Grassland

The correct answer is A:) Taiga. Taiga is the Russian word for forest. A taiga is frozen at least six months out of the year. Summers are very short, with only 50 to 100 days without frost.

63) What is produced in fermentation?

   A) Carbon dioxide
   B) 4 ATP
   C) Lactate
   D) 2 ATP and toxins
   E) None of the above

The correct answer is D:) 2 ATP and toxins. Fermentation is an anaerobic process, which produces a mere 2 ATP and toxins. It can be useful however, when short bursts of energy are needed.

64) In an angiosperm, how many microspore mother cells are required to produce 100 pollen grains?

   A) 25
   B) 100
   C) 50
   D) 75
   E) None of the above

The correct answer is A:) 25.

65) What causes a plant to bend towards sunlight?

   A) Abscistic acid
   B) Gibberellins
   C) Phototropin
   D) Auxin
   E) Cytokinins

The correct answer is D:) Auxin. Auxin gathers in the shaded side of the stem and causes the cells to elongate.

66) A tertiary consumer is:

   A) Carnivore that eats another carnivore
   B) Animal that eats plants
   C) Primary producer
   D) Where transpiration takes place
   E) None of the above

The correct answer is A:) Carnivore that eats another carnivore.

67) Which of the following is NOT a nucleotide used in DNA?

   A) Adenine
   B) Guanine
   C) Cytosine
   D) Thymine
   E) All of the above are nucleotides

The correct answer is E:) All of the above are nucleotides. Answers A through D list the four types of nucleotides.

68) Where the eggs are developed

   A) Ovary
   B) Ovule
   C) Stamen
   D) Pistil
   E) None of the above

The correct answer is A:) Ovary.

69) Which of the following describes conjugation?

   A) When bacteria picks up excess DNA from their surroundings, released by prokaryotic organisms.
   B) When bacteriophages carry bacterial DNA from one cell to another.
   C) The removal of dangerous pollutants from an area, using bacteria to absorb the pollutants.
   D) The transfer of DNA, from a donor cell to a recipient cell, through a temporary joint between the two.
   E) When a bacteria sends enzymes into the environment to decompose organic molecules.

The correct answer is D:) The transfer of DNA, from a donor cell to a recipient cell, through a temporary joint between the two. Answer A describes transformation and answer B describes transduction, both of which are also ways DNA is changed in bacteria.

70) What are nucleic acids also called?

   A) Nucleic secretions
   B) DNA and RNA
   C) Ribosomes
   D) Nucleotides
   E) All of the above

The correct answer is B:) DNA and RNA. DNA and RNA are called nucleic acids because they were first discovered in the nucleus of cells. Nucleotides are a major component of nucleic acids, but not another name for them.

71) Which organelle creates lipids?

   A) The endoplasmic reticulum
   B) The mitochondrion
   C) The ribosome
   D) The Golgi complex
   E) The nucleus

The correct answer is A:) The endoplasmic reticulum. The endoplasmic reticulum packages proteins before they travel to the Golgi complex.

72) Azotobacter and Bacillus polymyxa are the examples of

   A) Decomposers
   B) Non-symbiotic N2 fixer
   C) Symbiotic N2 fixer
   D) Pathogenic bacteria
   E) None of the above

The correct answer is B:) Non-symbiotic N2 fixer.

73) In C4 plants, CO, combines with

   A) Phosphoglycer aldehyde
   B) Phosphoenol pyruvate
   C) Phosphoglyveric acid
   D) Ribulose phosphate
   E) None of the above

The correct answer is B:) Phosphoenol pyruvate.

74) In which ecosystem would a great diversity of evergreen broadleaved trees, vines and epiphytes be expected?

   A) Prairie
   B) North American grassland
   C) Tropical rain forest
   D) Desert
   E) Temperate coniferous forest

The correct answer is C:) Tropical rain forest. An average tropical rain forest receives 50 to 260 inches of rain annually and the temperatures range from 68 to 90 degrees Fahrenheit.

75) A gymnospermic leaf carries 16 chromosomes. The number of chromosomes in its endosperm will be

   A) 12
   B) 24
   C) 8
   D) 4
   E) None of the above

The correct answer is C:) 8.

76) Which of the following correctly describes why it is necessary to administer booster injections of a vaccine?

   I. B cells wear out over long periods of time and need to be replenished.
   II. If a person doesn't have booster shots, they are actually MORE likely to become ill.
   III. If a virus mutates, then a new vaccination is needed which is designed to protect against it.

   A) I only
   B) I and II only
   C) II and III only
   D) I and III only
   E) I, II and III

The correct answer is D:) I and III only. I and III are both reasons why booster vaccinations are necessary. However, it is not true that not getting a booster shot increases a person's chance of becoming ill.

77) The enzyme enterokinase helps in the conversion of

   A) Caesinogen into caesin
   B) Pepsinogen into pepsin
   C) Proteins into polypeptides
   D) Trypsinogen into trypsin
   E) None of the above

The correct answer is D:) Trypsinogen into trypsin.

78) Besides annelida and arthropoda, the metamerism is exhibited by

   A) Acanthocephala
   B) Chordate
   C) Cestoda
   D) Mollusca
   E) None of the above

The correct answer is B:) Chordate.

79) The function of contractile vacuole in protozoa is

   A) Reproduction
   B) Locomotion
   C) Osmoregulation
   D) Digestion of food
   E) None of the above

The correct answer is C:) Osmoregulation.

80) Which of the following is NOT true of niches?

   A) A niche includes the habitat of a species.
   B) The resources in a habitat play a role in determining an animal's niche.
   C) A species interactions with other species factor into their niches.
   D) Every organism in an environment has a niche.
   E) Often there are two species which belong to each niche.

The correct answer is E:) Often there are two species which belong to each niche. The purpose of a niche is that it is the unique place a species fits in a habitat.

81) The hemorrhagic disease of newborns is caused due to the deficiency of

   A) Vitamin Bl2
   B) Vitamin A
   C) Vitamin K
   D) Vitamin B1
   E) None of the above

The correct answer is C:) Vitamin K.

82) Which of the following is TRUE of the lowest level of the food chain?

   A) The smallest amount of stored energy is found there.
   B) The fewest species are contained because many species are eaten to extinction.
   C) The largest amount of stored energy is found there.
   D) It represented the least amount of biomass.
   E) None of the above are true statements

The correct answer is C:) The largest amount of stored energy is found there. This is explained by the biomass pyramid. It is necessary that the lower levels contain more biomass to support higher up levels.

83) According to the accepted concept of hormone action, if receptor molecules are removed from target organs, then the target organ will

   A) Continue to respond to the hormone without any difference.
   B) Continue to respond to the hormone but will require higher concentration.
   C) Continue to respond to the hormone but in the opposite way.
   D) Not respond to the hormone.
   E) None of the above

The correct answer is D:) Not respond to the hormone.

84) The mammalian corpus luteum produces

   A) Luteinizing hormone
   B) Progesteron
   C) Estrogen
   D) Luteotropic hormone
   E) None of the above

The correct answer is B:) Progesteron.

85) What do ribosomes do?

   A) Communicates and give instructions to all other parts of the cell.
   B) Transport information from one organelle to another.
   C) Enclose and protect the cell from outside influence.
   D) Synthesize proteins for the cell to use.
   E) Aid in the condensing of DNA, so that it will fit within the nuclear membrane.

The correct answer is D:) Synthesize proteins for the cell to use. They are able to do this along with help from mRNA and rRNA.

86) The neurogenic heart characteristic feature of

   A) Humans
   B) Rabbits
   C) Rats
   D) Lower vertebrates
   E) None of the above

The correct answer is D:) Lower vertebrates.

87) Which of the following does NOT belong to the phylum Chordata?

   A) Fish
   B) Crabs
   C) Birds
   D) Dogs
   E) Lizards

The correct answer is B:) Crabs. The phylum Chordata contains vertebrate animals, and crabs belong to the phylum Anthrapoda. All of the other options do belong to the phylum Chordata.

88) The vitamin C or ascorbic acid prevents

   A) Antibody synthesis
   B) Rickets
   C) Scurvy
   D) Pellagra
   E) None of the above

The correct answer is C:) Scurvy.

89) What is the cell membrane primarily composed of?

   A) Carbohydrates
   B) Fatty Acids
   C) Phospholipid bilayer
   D) Proteins
   E) Specially bonded enzymes

The correct answer is C:) Phospholipid bilayer. Because the membrane is selectively permeable, the cell is still able to transport through the membrane.

90) In veins, valves are present to check backward flow of blood flowing at

   A) Atmospheric pressure
   B) Low pressure
   C) High pressure
   D) All these
   E) None of the above

The correct answer is B:) Low pressure.

91) Diapedesis is

   A) Formation of WBC
   B) Formation of pus
   C) Bursting of WBC
   D) Passage of WBC
   E) None of the above

The correct answer is D:) Passage of WBC.

92) Stomach in vertebrates is the chief site for digestion of

   A) Carbohydrates
   B) Proteins
   C) Fats
   D) All of the above
   E) None of the above

The correct answer is B:) Proteins.

93) What is mutualism?

   A) A symbiotic relationship which both members are harmed by.
   B) A symbiotic relationship which both members benefit from.
   C) A symbiotic relationship which benefits one member and harms the other.
   D) A symbiotic relationship which harms one member and doesn't affect the other.
   E) A symbiotic relationship which benefits one member and does affect the other.

The correct answer is B:) A symbiotic relationship which both members benefit from. Mutualism is recognized as highly important in shaping community structure.

94) What is the function of enzymes?

   A) Make up, hair and fingernails, as well as support skin ligaments and tendons.
   B) Allow substances to transport through cell membranes.
   C) Stop disease causing cells from upsetting homeostasis.
   D) To speed up specific chemical reactions within cells.
   E) Enzymes carry messages within the cells to cause movement.

The correct answer is D:) To speed up specific chemical reactions within cells. Enzymes are a type of protein and they bring reactants together.

95) Intervertebral disc is made up of

   A) Hyaline cartilage
   B) Fibrous cartilage
   C) Elastic cartilage
   D) Calcified cartilage
   E) None of the above

The correct answer is B:) Fibrous cartilage.

96) Normally the genes for antibiotic resistance are found in (in bacteria)

   A) Nucleus
   B) Plastid
   C) Chromosome
   D) Cell wall
   E) None of the above

The correct answer is B:) Plastid.

97) The scientists famous for chromosome heredity are

   A) Beadle and Tatum
   B) Morgan and Bridges
   C) Sutton and Boveri
   D) Davenport and Mendel
   E) None of the above

The correct answer is C:) Sutton and Boveri.

98) The science dealing with the improvement of human race by controlled selective crossing with individuals with desirable characteristics is called

   A) Euthenics
   B) Genetics
   C) Heterosis
   D) Eugenics
   E) None of the above

The correct answer is D:) Eugenics.

99) In Drosophila (fruit fly), there are

   A) Two linkage group
   B) Four linkage group
   C) Six linkage group
   D) Eight linkage group
   E) None of the above

The correct answer is D:) Eight linkage group.

100) Which of the following correctly states the path of urine after reaching the kidney?

   A) Urethra, ureters, bladder
   B) Bladder, ureters, urethra
   C) Bladder, urethra, ureters
   D) Ureters, bladder, urethra
   E) Ureters, urethra, bladder

The correct answer is D:) Ureters, bladder, urethra. The ureters connect the kidney and the bladder, where the urine waits until exiting the body through the urethra.

101) The condition when the chromosome number increases or decreases, may be due to

   A) Genetic drift
   B) Mutagenic change
   C) Non-disjunction of chromosome
   D) Gene mapping
   E) None of the above

The correct answer is C:) Non-disjunction of chromosome.

102) Which one of the following is a sex-linked inheritance?

   A) Anemia
   B) Night blindness
   C) Colorblindness
   D) Meningitis
   E) None of the above

The correct answer is C:) Colorblindness.

103) The enzyme by which the recombinant DNA is achieved by cleaving the pro DNAs are

   A) Primases
   B) Ligase
   C) Exonucleases
   D) Restriction endonucleases
   E) None of the above

The correct answer is D:) Restriction endonucleases.

104) The process of exchange of chromatida between non-homologous chromosomes is called

   A) Duplication
   B) Illegitimate crossing over
   C) Deletion
   D) Inversion
   E) None of the above

The correct answer is B:) Illegitimate crossing over.

105) Which of the following correctly states the function of the hypothalamus?

    A) Regulate reflexive motions such as breathing and heartbeat.
    B) Receive sensory and motor input.
    C) Maintain homeostasis by regulating factors such as hunger, sleep and water balance.
    D) Secretes melatonin while a person is sleeping.
    E) To make the conscious mind aware of past experiences.

The correct answer is C:) Maintain homeostasis by regulating factors such as hunger, sleep and water balance. Answer A describes the medulla oblongata. Answer B describes the cerebellum. Answer D describes the pineal gland. Answer E describes the hippocampus.

106) Chromosomes and nucleus are particularly stained by carmine, a dye extracted from

    A) Berberis
    B) Azadirechta
    C) Coccus cocti
    D) Ficus
    E) None of the above

The correct answer is C:) Coccus cocti.

107) What is bending movement towards the light in plants called?

    A) Positive phototropism
    B) Gravitropism
    C) Auxinism
    D) Cotylotation
    E) Homeostasis

The correct answer is A:) Positive phototropism. This is caused by auxin in the shaded side of the stem.

108) The rediscovery of Mendelism was by

   A) De Vries, Collins, T. Schermack
   B) Boveri, Sutton, Tippo
   C) Davenport, Johansson, Jacob
   D) De Vries, Correns, T. Schermac
   E) None of the above

The correct answer is D:) De Vries, Correns, T. Schermac.

109) When a child or animal comes to see another as a parent or thing as an object of trust it is called

   A) Translation
   B) Habituation
   C) Imprinting
   D) Ficus
   E) None of the above

The correct answer is C:) Imprinting.

110) What is the function of organelles?

   A) To produce hair and fingernails and support tendons.
   B) To speed up specific chemical reactions within cells.
   C) To create white blood cells which fight disease.
   D) To carry out specific cell processes.
   E) All of the above

The correct answer is D:) To carry out specific cell processes. The reason cells can be efficient is because they make use of different organelles to perform different processes. Each different organelle is responsible for a different task.

111) If one strand of DNA has the nitrogenous base - GCTTGAAT, the sequence in the complementary strand will be

   A) CGACTTAT
   B) CGAACTAT
   C) CGAACTTA
   D) CGAACATT
   E) None of the above

The correct answer is C:) CGAACTTA.

112) How do enzymes affect the activation energy needed?

   A) Decrease it, because the purpose of enzymes is to bring reactants together, causing reactions to occur more quickly than if it occurred naturally.
   B) Decrease it, because the purpose of enzymes is to bring reactants together, causing reactions to occur more slowly than if it occurred naturally.
   C) Increase it, because the purpose of enzymes is to bring reactants together causing reactions to occur more slowly than if it occurred naturally.
   D) Increase it, because the purpose of enzymes is to bring reactants together, causing reactions to occur more quickly than if it occurred naturally.
   E) All of the above, it depends upon the specific case.

The correct answer is A:) Decrease it, because the purpose of enzymes is to bring reactants together causing reactions to occur more quickly than if it occurred naturally. Answer B gives the correct response, but the incorrect explanation. Answer D gives the correct explanation, but the incorrect response.

113) The pneumonia causing bacteria - Pneumonococcus experiment proves that

   A) RNA controls the production of DNA and proteins
   B) DNA is the genetic material
   C) Bacteria do not show binary fission
   D) Bacteria do not produce by sexual method
   E) None of the above

The correct answer is B:) DNA is the genetic material.

114) The process in which DNA of a bacterial cell is transferred into another bacterial cell by a virus is called

   A) Transformation
   B) Translation
   C) Transcription
   D) Transduction
   E) None of the above

The correct answer is D:) Transduction.

115) The book published by Jean Baptise de Lamarck, a French biochemist, was

   A) Philosophic Zoologique
   B) Theory of Natural Selection
   C) Descent with modification
   D) Origin of Species
   E) None of the above

The correct answer is A:) Philosophic Zoologique.

116) What is another name for oxygen isotopes 17O and 18O?

   A) Heavy oxygen
   B) Unstable oxygen
   C) High oxygen
   D) Complete oxygen
   E) None of the above

The correct answer is A:) Heavy oxygen. Heavy oxygen makes up about .2% of natural oxygen.

117) A total sum of genes present in all the individuals in a population is called

   A) Genome
   B) Genotype
   C) Gene pool
   D) Clone
   E) None of the above

The correct answer is C:) Gene pool.

118) The phenomenon that prevents member of a species from interbreeding and preserves the integrity of a species by not allowing hybridization is called

   A) Variation
   B) Speciation
   C) Reproductive isolation
   D) Genetic isolation
   E) None of the above

The correct answer is C:) Reproductive isolation.

119) If both parents carry the sickle cell trait, what is the chance their child will have sickle cell disease?

   A) 10%
   B) 25%
   C) 50%
   D) 75%
   E) 90%

The correct answer is B:) 25%. Because each parent is a carrier, they both must have one copy of the gene. The heterozygous cross will yield a 25% chance of two recessive genes occurring.

120) The metal tolerant plants which serve as indicators of the occurrence of specific metal deposits are known as

   A) Bio-medical plants
   B) Bio intimation plants
   C) Bio-signal plants
   D) Bio-indicator plants
   E) None of the above

The correct answer is D:) Bio-indicator plants

121) The scientist(s) giving the new concept of Neo-Darwinism was/were

   A) Weismann
   B) De Veries
   C) Stanley Miller
   D) Gold Schmidt and Dob Zehnsky
   E) None of the above

The correct answer is D:) Gold Schmidt and Dob Zehnsky.

122) What type of cell is responsible for detecting antigens and producing antibodies?

   A) Helper T cells
   B) Cytotoxic T cells
   C) B-Lymphocytes
   D) Memory T cells
   E) None of the above

The correct answer is C:) B-Lymphocytes. Thousands of B-Lymphocytes, or B cells, are produced in the bone marrow every day.

123) A mule is the hybrid produced by a cross breeding between a

   A) Male horse and female donkey
   B) Male donkey and female horse
   C) Male horse and female cat
   D) Male monkey and female dog
   E) None of the above

The correct answer is B:) Male donkey and female horse.

124) Biston-betularia (light colored peppered moth) changed to darker variety (Biston carbonaria) due to

   A) Mutation
   B) Genetic isolation
   C) Reproductive isolation
   D) Normal change
   E) None of the above

The correct answer is A:) Mutation.

125) Where are osteoblasts found?

   A) Fingernails
   B) Muscles
   C) Organs
   D) Bones
   E) Skin Cells

The correct answer is D:) Bones. Osteoblasts are cells which form the bone matrix by laying down hard mineral material.

126) The scientist who worked with Darwin for providing the concept of origin of species by natural selection was

   A) Prof. Y.D. Tyagi
   B) Weismann
   C) Alfred Wallace
   D) Ledenberg
   E) None of the above

The correct answer is C:) Alfred Wallace.

127) According to Darwin regarding evolution, which one of the following is correct?

    A) Original species have a greater chance of evolution
    B) Almighty God determines the evolution of any species
    C) Progressive adaptations provide the ability to one species to have more progeny
    D) Any species may be evolved at any time
    E) None of the above

The correct answer is C:) Progressive adaptations provide the ability to one species to have more progeny.

128) Adaptation of different types of beaks of finches to different types of feeding habituates on the Galapagos Islands show

    A) Intraspecific variation
    B) Interspecific competition
    C) Inheritance of acquired characters
    D) Origin of species by natural selection
    E) None of the above

The correct answer is D:) Origin of species by natural selection.

129) Which of the following is TRUE?

    A) The cell membrane is composed only of a phospholipid bilayer.
    B) Plant cells only have a cell membrane, whereas animal cells have a cell membrane and a cell wall.
    C) The cell membrane is selectively permeable.
    D) Cytoplasm exists only in plant cells.
    E) All of the above are true statements.

The correct answer is C:) The cell membrane is selectively permeable. This means that the cell can regulate what is let through the membrane.

130) Succession may be defined as

    A) Gradual convergent directional and continuous process
    B) Series of biotic communities that appear gradually in a barren area
    C) Redistribution of genes among a population
    D) Both A and B
    E) None of the above

The correct answer is D:) Both A and B.

131) All vertebrates have gill slits in their embryonal stage which shows

   A) Recapitulation
   B) Metamorphosis
   C) Biogenesis
   D) Biogenetic law
   E) None of the above

The correct answer is C:) Biogenesis.

132) Struggle among the individuals of different species having variable needs is called

   A) Intraspecific struggle
   B) Interspecific struggle
   C) Environmental struggle
   D) Normal struggle
   E) None of the above

The correct answer is B:) Interspecific struggle.

133) Which of the following is NOT a common element of plant and animal cells?

   A) Cell wall
   B) Cytoplasm
   C) Mitocondria
   D) Golgi complex
   E) Nucleus

The correct answer is A:) Cell wall. Animal cells only have a cell membrane; however, plant cells have a cell membrane and a cell wall.

134) The scientist who discovered population of Impatiens-balasamina growing around Zwar Zinc mines of Udaipur is

   A) Charles Darwin
   B) Prof. P.K. Sethi
   C) Prof. Y.D. Tyagi
   D) Dr. M.S. Swaminathan
   E) None of the above

The correct answer is C:) Prof. Y.D. Tyagi.

135) The changes in gene frequencies due to chance effect are collectively called

   A) Genetic drift
   B) Genetic isolation
   C) Genome change
   D) Genotypic variation
   E) None of the above

The correct answer is A:) Genetic drift.

136) In Down's syndrome, Karyotyping has shown that the disorder is associated with trisomy of chromosome number-21 usually due to

   A) Non-disjunction during formation of egg-cells and sperm-cells
   B) Non-disjunction during egg-cell formation
   C) Addition of extra chromosome during mitosis of the zygote
   D) Non-disjunction during sperm-cell formation
   E) None of the above

The correct answer is B:) Non-disjunction during egg-cell formation.

137) Mental retardation in man associated with sex chromosomal abnormality is usually due to

   A) Large increase in Y complement
   B) Increase in X complement
   C) Reduction in X complement
   D) Moderate increase in Y complement
   E) None of the above

The correct answer is B:) Increase in X complement.

138) Do both DNA and RNA have a phosphate group?

   A) Yes, because both are nucleic acids, which contain phosphate groups.
   B) Yes, because the phosphate group of DNA is deoxyribose and the phosphate group of RNA is ribose.
   C) No, because only RNA has a phosphate group.
   D) No, because only DNA has a phosphate group.
   E) None of the above correctly answer the question.

The correct answer is A:) Yes, because both are nucleic acids, which contain phosphate groups. Answer B is incorrect because deoxyribose and ribose are sugars, not phosphate groups.

139) When an infant no longer responds to a new toy he is showing

   A) Habituation
   B) Object permanence
   C) Hyperactivity
   D) Divergence
   E) Convergence

The correct answer is A:) Habituation.

140) Where in the cell are the ribosomes located?

   A) Cell membrane
   B) Nucleus
   C) Cytoplasm
   D) Golgi apparatus
   E) Mitochondria

The correct answer is C:) Cytoplasm. There can be anywhere from thousands to millions of ribosomes in any given cell.

141) Hybridoma cells are

   A) Product of spore formation in bacteria
   B) Hybrid cells resulting from myeloma cells
   C) Nervous cells of frog
   D) Only cells having oncogenes
   E) None of the above

The correct answer is B:) Hybrid cells resulting from myeloma cells.

142) The aquatic fern, which is an excellent bio-fertilizer is

   A) Azolla
   B) Pteridium
   C) Marsilia
   D) Terlangier
   E) None of the above

The correct answer is A:) Azolla.

143) How is gender determined in humans?

   A) Number of X chromosomes
   B) Total number of chromosomes
   C) Amount of sex chromosomes only
   D) Presence or absence of Y chromosome
   E) None of the above

The correct answer is D:) Presence or absence of the Y chromosome. This is why it is possible for people to live fairly normal lives when born with excess X chromosomes.

144) The new varieties of plants are produced by

   A) Selection and hybridization
   B) Selection and introduction
   C) Mutation and selection
   D) Introduction and mutation
   E) None of the above

The correct answer is A:) Selection and hybridization.

145) Due to which of the following organisms has the yield of rice been increased?

   A) Anabaena
   B) Bacillus polymyxa
   C) Bacillus papilliae
   D) Sesbania
   E) None of the above

The correct answer is A:) Anabaena.

146) Which of the following is NOT a component of nucleotides?

   A) A phosphate
   B) A nitrogen base
   C) A 5-carbon sugar called deoxyribose
   D) All of the above
   E) None of the above

The correct answer is D:) All of the above. Answers A, B and C list the three components of a nucleotide.

147) The first transgenic crop was

   A) Cotton
   B) Pea
   C) Tobacco
   D) Flax
   E) None of the above

The correct answer is D:) Flax.

148) What is deoxyribose?

   A) A phosphate group which supports DNA production.
   B) A nucleotide which builds the basic structure of DNA.
   C) A nitrogen containing base which is essential to DNA development.
   D) A sugar which is a component of DNA.
   E) A chemical which causes the double helix shape of DNA strands.

The correct answer is D:) A sugar which is a component of DNA. Deoxyribose is where DNA gets its name, deoxyribonucleic acid.

149) The problem due to Rh-factor arises when the blood of two (Rh+ and Rh-) mix-up

   A) During pregnancy
   B) Through transfusion
   C) In a test tube
   D) Both A and B
   E) None of the above

The correct answer is A:) During pregnancy.

150) The fish that eradicate the mosquito larva are known as

   A) Cutter fish
   B) Anabas
   C) Gambusia
   D) Rohu
   E) None of the above

The correct answer is C:) Gambusia.

151) Hominidae is part of which group?

   A) Kingdom
   B) Phylum
   C) Order
   D) Family
   E) Genus

The correct answer is D:) Family.

152) The blood group with antibody-A and antibody-B is

   A) B
   B) A
   C) O
   D) AB
   E) None of the above

The correct answer is D:) AB.

153) The opening in the leaves where transpiration takes place is called

   A) Roots
   B) Chloroplast
   C) Stoma
   D) Xylem
   E) Phloem

The correct answer is C:) Stoma.

154) Which of the following is NOT an ideal condition for the Hardy-Weinberg Law?

   A) Large population size
   B) Mutation
   C) No immigration or emigration
   D) Random mating
   E) Random reproductive success

The correct answer is B:) Mutation. The ideal conditions include large population size, no mutation, no immigration or emigration, random mating and random reproductive success.

155) Which of the following correctly describes the locations of the Calvin and Krebs cycles respectively?

A) The stroma and mitochondria of both plants and animals.
B) The stroma of plants, the mitochondria of plants and animals.
C) The stroma of plants and animals, the mitochondria of plants.
D) The mitochondria of plants and animals, the stroma of plants.
E) The mitochondria of plants, the stroma of plants and animals.

The correct answer is B:) The stroma of plants, the mitochondria of plants and animals. The Calvin cycle occurs in the stroma of photosynthetic plants, and the Krebs cycle occurs in the mitochondria of plants and animals.

156) The term humulin is used for

A) Human insulin
B) Powerful antibiotic
C) Isoenzyme
D) Hydrolytic enzyme
E) None of the above

The correct answer is A:) Human insulin.

157) Water is a "polar" molecule, meaning that one side is positively charged and one side is negatively charged. Water is polar because

A) Electrons are more drawn to the oxygen than hydrogen, causing the hydrogen atom to have a negative charge.
B) Hydrogen bonds are stronger than covalent bonds, giving water its unique characteristics.
C) The atoms have a unique covalent bond because of their shared protons.
D) Because of its shape, the oxygen atoms can bond with the hydrogen atoms of several other water molecules at once.
E) None of the above

The correct answer is D:) Because of its shape, the oxygen atoms can bond with the hydrogen atoms of several other water molecules at once.

158) Which of the following is NOT a type of selection?

   A) Directional
   B) Diversifying
   C) Stabilizing
   D) Environmental
   E) All of the above are types of selection

The correct answer is D:) Environmental.

159) The terminal codons are (or terminating codon)

   A) UAA, UAG, UGA
   B) UAA, AUG, GUG
   C) GUG, AUG, UGA
   D) UGA, UAA, AUG
   E) None of the above

The correct answer is A:) UAA, UAG, UGA.

160) Which of the following is NOT true about sclereids?

   A) These are found in nut shells, guava pulp and pears
   B) These are also called stone cells
   C) These are groups of living cells
   D) These are a form of sclerenchyma with fibers
   E) None of the above

The correct answer is C:) These are groups of living cells.

161) During photosynthesis, oxygen in glucose comes from

   A) Water
   B) Oxygen in air
   C) Carbon dioxide
   D) Both A and B
   E) All of the above

The correct answer is C:) Carbon dioxide.

162) The tubes of a plant that transport the nutrients are known as

    A) Xylem
    B) Phloem
    C) Carbon dioxide
    D) Both A and B
    E) None of the above

The correct answer is A:) Xylem.

163) The thickening of walls of arteries are called

    A) Arthritis
    B) Arteriosclerosis
    C) Aneurysm
    D) Both A and C
    E) None of the above

The correct answer is B:) Arteriosclerosis.

164) Sound can be transmitted in all of the following except

    A) Air
    B) Water
    C) A diamond
    D) A vacuum
    E) None of the above

The correct answer is D:) A vacuum. Sound, a longitudinal wave, is transmitted by vibrations of molecules, so it can be transmitted through any gas, liquid or solid. But it cannot be transmitted through a vacuum, because there are no particles present to vibrate and bump into their adjacent particles to transmit the waves.

165) In a certain population of 1000 tree frogs, 640 have red eyes while the rest have sepia eyes. The sepia eye trait is recessive to red eyes. How many individuals in this population would you expect to be homozygous for red eye color?

    A) 10
    B) 60
    C) 160
    D) 400
    E) None of the above

The correct answer is C:) 160.

166) The Minamata disease in Japan was caused through pollution of water by

   A) Methyl isocyanate
   B) Mercury
   C) Lead
   D) Cyanide
   E) None of the above

The correct answer is B:) Mercury.

167) DDT is

   A) A non-degradable pollutant
   B) An antibiotic
   C) A biodegradable pollutant
   D) Not a pollutant
   E) None of the above

The correct answer is A:) A non-degradable pollutant.

168) Which of the following is the least inclusive group?

   A) Kingdom
   B) Phylum
   C) Order
   D) Family
   E) Species

The correct answer is E:) Species.

169) Volume and surface area of the cell have

   A) No relation
   B) Occasional relation
   C) Direct proportion relation
   D) Inverse proportion relation
   E) None of the above

The correct answer is D:) Inverse proportion relation.

170) Size of the cell is governed by

A) Cell volume and cell surface area ratio
B) Nucleus and cytoplasm ratio
C) Rate of cellular metabolism
D) All the above
E) None of the above

The correct answer is A:) Cell volume and cell surface area ratio.

171) The closure of the lid of the pitcher in the pitcher plant is due to

A) Autonomous movement
B) Turgor movement
C) Paratonic movement
D) Tropic movement
E) None of the above

The correct answer is C:) Paratonic movement.

172) What type of cells does Human Immunodeficiency Virus (HIV) attack?

A) Helper T cells
B) Cytotoxic T cells
C) B-Lymphocytes
D) Memory T cells
E) None of the above

The correct answer is A:) Helper T cells. Helper T cells are important because they signal the B cells to make antibodies and activate the cytotoxic T cells.

173) Coir is the commercial product of coconut's

A) Endosperm
B) Mesocarp
C) Pericarp
D) Endocarp
E) None of the above

The correct answer is B:) Mesocarp.

# Test Taking Strategies

Here are some test-taking strategies that are specific to this test and to other CLEP tests in general:
- Keep your eyes on the time. Pay attention to how much time you have left.
- Read the entire question and read all the answers. Many questions are not as hard to answer as they may seem. Sometimes, a difficult sounding question really only is asking you how to read an accompanying chart. Chart and graph questions are on most CLEP tests and should be an easy free point.
- If you don't know the answer immediately, the new computer-based testing lets you mark questions and come back to them later if you have time.
- Read the wording carefully. Some words can give you hints to the right answer. There are no exceptions to an answer when there are words in the question such as "always" "all" or "none." If one of the answer choices includes most or some of the right answers, but not all, then that is not the answer. Here is an example:

    The primary colors include all of the following:

    A) Red, Yellow, Blue, Green
    B) Red, Green, Yellow
    C) Red, Orange, Yellow
    D) Red, Yellow, Blue
    E) None of the above

    Although item A includes all the right answers, it also includes an incorrect answer, making it incorrect. If you didn't read it carefully, were in a hurry, or didn't know the material well, you might fall for this.
- Make a guess on a question that you do not know the answer to. There is no penalty for an incorrect answer. Eliminate the answer choices that you know are incorrect. For example, this will let your guess be a 1 in 3 chance instead.

# What Your Score Means

Based on your score, you may, or may not, qualify for credit at your specific institution. At University of Phoenix, a score of 50 is passing for full credit. At Utah Valley State College, the score is unpublished, the school will accept credit on a case-by-case basis. Another school, Brigham Young University (BYU) does not accept CLEP credit.

To find out what score you need for credit, you need to get that information from your school's website or academic advisor.

You can score between 20 and 80 on any CLEP test. Some exams include percentile ranks. Each correct answer is worth one point. You do not lose points for unanswered or incorrect questions.

Test Preparation

How much you need to study depends on your knowledge of a subject area. If you are interested in literature, took it in school, or enjoy reading then your studying and preparation for the literature or humanities test will not need to be as intensive as someone who is new to literature.

This book is much different than the regular CLEP study guides. This book actually teaches you the information that you need to know to pass the test. If you are particularly interested in an area, or feel like you want more information, do a quick search online. We've tried not to include too much depth in areas that are not as essential on the test. Everything in this book will be on the test. It is important to understand all major theories and concepts listed in the table of contents. It is also very important to know any bolded words.

Don't worry if you do not understand or know a lot about the area. With minimal study, you can complete and pass the test.

To prepare for the test, make a series of goals. Allot a certain amount of time to review the information you have already studied and to learn additional material. Take notes as you study, it will help you learn the material.

Legal Note

All rights reserved. This Study Guide, Book and Flashcards are protected under US Copyright Law. No part of this book or study guide or flashcards may be reproduced, distributed or stored in a retrieval system, or transmitted in any form or by any means, electronic, mechanical, photocopying, recording, or otherwise, without the prior written permission of the publisher Breely Crush Publishing LLC. This manual is not supported by or affiliated with the College Board, creators of the CLEP test. CLEP is a registered trademark of the College Entrance Examination Board, which does not endorse this book.

References

[1] http://www.albinism.org/publications/what_is_albinism.html Reprinted with permission.
[2] http://www.ninds.nih.gov/health_and_medical/disorders/taysachs_doc.htm
The National Institute of Neurological Disorders and Stroke
National Institutes of Health
Bethesda, MD 20892
[3] U.S. DEPARTMENT OF HEALTH AND HUMAN SERVICES
Public Health Service
National Institutes of Health
National Heart, Lung, and Blood Institute
NIH Publication No. 95-3650
November 1995

http://www.nhlbi.nih.gov/health/public/lung/other/cf.htm

FLASHCARDS

This section contains flashcards for you to use to further your understanding of the material and test yourself on important concepts, names or dates. Read the term or question then flip the page over to check the answer on the back. Keep in mind that this information may not be covered in the text of the study guide. Take your time to study the flashcards, you will need to know and understand these concepts to pass the test.

Photosynthesis	Chlorophyll does what?
In photosynthesis, water and carbon dioxide is changed in what?	Plants create what?
Stomata	What opens to let in air and release oxygen in photosynthesis?
Transpiration	Plants need what to make food?

Absorbs sunlight	How plants take energy from the sun and turn it into food
Sugar and Starches	Oxygen and Glucose
Stomata	Small openings on the bottom of leaves
Water and carbon dioxide	Evaporation of water from the plants when the stomata is open during photosynthesis

Primary producer	Herbivores
Carnivores	Primary consumer
Secondary consumers	Tertiary consumer
Eating a plant breaks what?	Chemical process from eating food results in what?

Animals that eat plants	Plants
Anything that eats plants	Meat eaters
Anything that eats a Secondary consumer	Anything that eats a Primary consumer
Free energy	The energy bond

Taxonomy	Monera
Protista	Fungi
Plantae	Animalia
There are more what then any other animal?	The scientific classification order for organisms is?

Single cell bacteria	Science of naming organisms
Mushrooms	Multi cell organisms, like Amoebae
Animals	Plants
1. Kingdom, 2. Phylum or Division, 3. Class, 4. Order, 5. Family, 6. Genus, 7. Species	Insects

For Humans, what is the complete classification?	Cordata
Phylogenic Tree	Homologous structures
Ecosystem	Sun produces what amount of energy
What percentage of the sun is used in photosynthesis?	Decomposer

Animals with a spinal cord	Animalia, Chordata, Mammalia, Primates, Hominidae, Sapiens
Similar bone structures	A tool that predicts probable evolution
Most	The most inclusive ecological area, includes air, water and energy
A fungus or bacteria that absorbs nutrients	0.012

Biome	**List some major biomes**
Name Charles Darwin's Ship	**What bird helped Darwin realize natural selection?**
Natural Selection	**Gregor Mendel**
In genetics, Upper case letters mean what? (ex. PP)	**Chromosomes come in what?**

Tropical rain forest, forest, desert, tundra, grassland, chaparral	A section of the ecosystem, defined by plant life and location
Finch	HMS Beagle
Father of Genetics	When nature determines what will evolve and what will not survive
Pairs	Upper case means dominant lower case means recessive

Chromosomes carry what?	Somatic Cell
How many chromosomes to a set?	Cloning
Gamettes	How many gametes do you need for reproduction?
Zygote	Hemophilia

Full set of chromosomes	Genes
Reproducing using just the somatic cell	46
2	Reproductive cells
Blood clotting disorder	Two fused gametes, the part of a human

Disorders that can be passed on	Polygenetic
Rh Factor	Albinism
Tay-sachs disease	Cystic Fibrosis
Sickle Cell Anemia	Amniocentesis

Genes that determine what is passed on	Colorblindness, Hemophilia, Muscular Dystrophy
People with little or no pigment to skin, eyes, and hair	Presence or absence of protein
Genetic disorder	Genetic disorder
Test of amniotic fluid	Abnormal red blood cell that makes clogs

Genetic engineering	Artificial insemination
Artificial Implant	Surrogate Mother
In Vitro Fertilization	Ovary
Ovulation	Where do sperm and egg meet?

Sperm is taken from the male and injected into the woman for conception	Adding desirable traits by inserting genes
Someone who carries an embryo who is unrelated	Zygote is inserted in a host body
Where eggs are developed	Fertilization done in a lab where the sperm and egg are combined then implanted
Fallopian tube	When the egg matures and breaks through the ovary

Birth control pills signal what to the body?	Yolk sac
Amniotic Cavity	Placenta
Allantois	Cells reproduce what?
Chromosomes act as a template in what?	DNA stands for what?

Sac filled with food for the embryo	That an egg has not attached and the lining is expelled
Placenta is the excretion track for the embryo	The area in the body where the baby develops
Themselves	The process of hold excretion
Deoxyribonucleic Acid	Cell reproduction

DNA molecules consist of how many chains?	How many pairs of chromosomes determine sex?
How many pairs are there in a chromosome?	How many total chromosomes is there in a nucleus?
Ribosome	What percentage of the body is water?
Carbohydrates are	Proteins are

One, X & Y	Two chains
46	23
0.9	Organelle
Sources of essential amino acids	Sugar and starches

Fats	Scurvy
Yeast is what?	Louis Pasteur
Malaria	DDT
Virus	Virus examples

Lack of vitamin C	Storage area for food
Father of Microbiology	Alive and an animal
Chemical used to kill mosquitoes, harmful to humans	Disease spread mostly by mosquitoes
Colds, AIDS, Herpes, Flu, Measles	Protein that mimics RNA and DNA

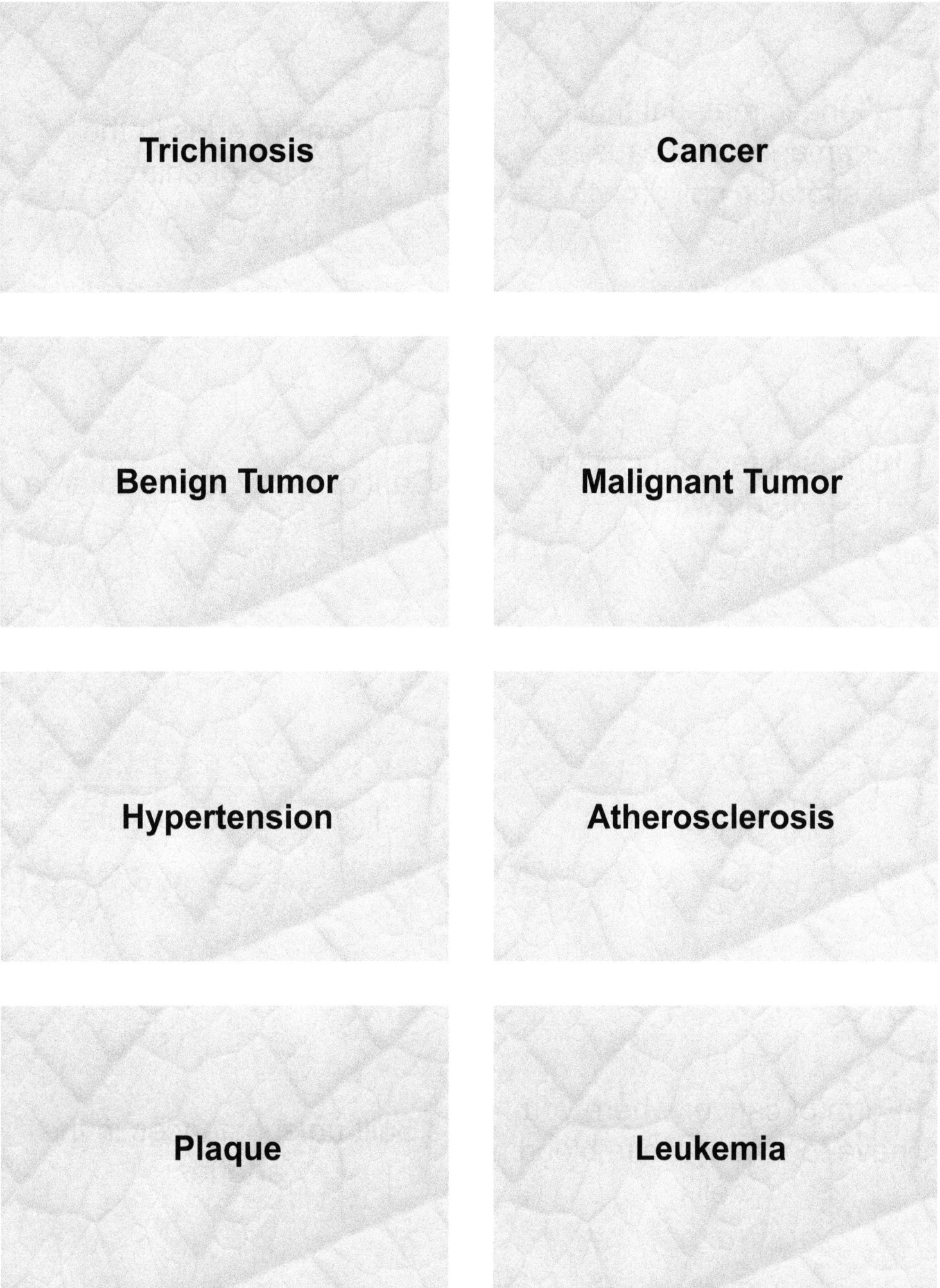

Genetic material that is damaged and causes sporadic cell growth	Parasite eggs in the muscles of animals
Unrestricted Cancer cell growth	Cancer restricted to an area
Hardened arteries	High Blood Pressure
Form of cancer where you have too many white blood cells	Built up substances in the arteries

Exhale	Thyroid regulates what?
What glands give the "Fight or Flight" response?	Leukocytes
Eurethrocytes	Four Major Muscle Tissue Types
Three Types of Muscle Tissue	Smooth Muscle examples

Metabolism	Breathing out carbon dioxide
White blood cells that fight infection	Adrenal glands
Epithelial, Connectiver, Nervous, Muscle	Red blood cells
Veins and arteries	Smooth, Skeletal, Cardiac

Skeletal Muscle examples	Cardiac
Hepatic Portal Circulation	Urea
Uric Acid	Ecotherms
Endotherms	Warm Blooded

Heart muscle	Striated, around skeleton
Nitrogenous waste of humans	Regulates sugar levels
Cold blooded animals	Paste like feces of birds and reptiles
Animals who regulate their own body temperature	Warm blooded animals

Cold Blooded	Endocrine System
Cornea	Echolocation
Pheromones	Social Insect Example
Which cell (animal or plant) has a cell wall?	Vascular Cambium

Glands that produce hormones for major glands	Animals who need a warm climate because they do not regulate their body temperature
Ultrasound that animals (bats) use to navigate	Protective covering of the eye
Bees, ants	Scents that insects use to communicate with each other
A structure on a plant that goes from the roots up to the veins of the leaves	Plants

Xylem	Stoma
Trees grow from the top or bottom up?	Stigma
Style	Anther
Stamen	Pistil

Openings in the leaves	Tree sap
Traps pollen grains	Top up
Produces pollen grains	Tube leading to plant ovary
Female plant organ	Male plant organ

www.ingramcontent.com/pod-product-compliance
Lightning Source LLC
Chambersburg PA
CBHW081830300426
44116CB00014B/2530